Peripheral Biological Markers in Alcoholism

Peripheral Biological Markers in Alcoholism

Nela Pivac, Dragica Kozarić-Kovačić, Maja Mustapić, Martina Deželjin, Korona Nenadić-Šviglin and Dorotea Muck-Šeler

Nova Biomedical Books
New York

For permission to use material from this book please contact us:
Telephone 631-231-7269; Fax 631-231-8175
Web Site: http://www.novapublishers.com

Library of Congress Cataloging-in-Publication Data

Peripheral biological markers in alcoholism / Nela Pivac ... [et al.] (editor).
p. ; cm.
Includes bibliographical references and index.
ISBN 978-1-60456-617-8 (softcover)
1. Alcoholism--Molecular aspects. 2. Biochemical markers. I. Pivac, Nela.
[DNLM: 1. Alcoholism. 2. Biological Markers--blood. WM 274 P445 2008]
RC565.P374 2008
616.86'1--dc22

2008017524

Published by Nova Science Publishers, Inc. New York

Contents

Preface

Alcoholism is a complicated disorder that is often characterized with refractoriness to treatment, chronic course, high comorbidity and frequently with aggressive and suicidal behavior. The neurobiological basis of alcoholism is assumed to be associated with the altered functions of different neurotransmitter (opioid, GABA, dopamine or serotonin (5-hydroxytrypta-mine, 5-HT)) systems. The association between central serotonergic function and the development and maintenance of alcohol abuse and dependence is based on the preclinical and clinical data, which show decreased 5-HT function associated with alcohol abuse. Peripheral biochemical markers might be used to improve the understanding of the underlying neurobiology of alcoholism, but also for the screening, establishing the diagnosis, disease staging, prediction of the suicidal risk and prediction and monitoring of treatment. Since 5-HT synaptosomes in the central nervous system and platelets similarly store, release and metabolize 5-HT, blood platelets have been widely used as a limited peripheral model for the central serotonergic synaptosomes. Various 5-HT abnormalities are frequently found in alcoholism, and decreased plasma tryptophan levels, decreased but also unaltered concentration of platelet 5-HT, increased platelet 5-HT uptake, and decreased and unchanged platelet monoamine oxidase (MAO) activity in alcoholic subjects have been reported. Deficits in central 5-HT function in alcoholic subjects were also detected after a challenge tests with fenflu-ramine, m-clorophenylpiperazine, and 6-chloro-2-1-piperazinylpyrazine, which resulted in blunted prolactin and cortisol responses. Alcoholism is frequently associated with different comorbid psychiatric disorders (personality disorder, PTSD, anxious-depressive disorder, major depression, schizophrenia), and with nicotine addiction.

Clinical diagnosis of alcoholism was made according to the DSM-IV criteria, using Structured clinical interview and CAGE. Different clinical symptoms were evaluated using Hamilton Rating Scale for Depression and Anxiety. Control group consisted of drug free healthy persons with no personal or family history of alcoholism, who did not consume any alcohol beverage 24 h prior to sampling. All subjects fulfilled the questionnaire regarding their smoking habits. Biomarkers studied were platelet 5-HT concentration, platelet MAO activity, and routine laboratory tests, determined using spectrofluorimetric and photometric methods.

Since platelet 5-HT concentration and/or MAO activity are under influence of sex, psychotic or depressive symptoms, suicidal behavior, smoking status, or different psychiatric comorbid diagnoses, we determined platelet 5-HT concentration and platelet MAO activity in the large groups of drug-free alcoholics who were admitted into hospital due to the withdrawal syndrome related to alcohol dependency, and compared these values in the sex-matched healthy controls. We also investigated whether sex, smoking status, the presence of psychotic or depressive symptoms, suicidal behavior, number of cigarettes smoked per day, or different psychiatric comorbid diagnoses, would affect platelet 5-HT and MAO in male and female alcoholic patients.

Our data suggest that biological markers are very important research tool in alcoholism, since they can be used to determine the baseline characteristics of the alcoholic groups, to predict a suicidal risk in alcoholism, to predict efficacy in alcoholism treatment trials, to predict therapeutic response to medications in alcoholism, to measure alcohol consumption, and to be used in forensic toxicology.

Keywords: biomarkers, platelets, serotonin, monoamine oxidase, symptoms, comorbid diagnoses, alcoholism.

Chapter 1

Introduction

1.1. Alcoholism

Alcoholism, including alcohol dependence, is a serious and global health problem, and according to the World Health Organization, it causes 3.2% of deaths (World Health Organization, 2002). In the developed countries alcoholism is a third leading course of the burden of disability-adjusted life years (World Health Organization, 2002). Alcoholism carries a great burden to patients and their families, and to a whole society, because of the disability and deaths. Prevalence of people with dependence or abuse of psychoactive substances (alcohol or illicit addictive drugs) in USA is, according to 2003 data from the National Survey on Drug Use and Health, around 9% of the total population (Substance Abuse and Mental Health Services Administration 2004). Alcoholism is characterized by the multiple pathological or biochemical abnormalities, dysfunctions in the neurotransmitter and neuromodulatory systems, with consequent characteristic clinical manifestations, that include excessive use of alcohol, impaired control of the alcohol consumption, tolerance to the effects of alcohol, and eventually withdrawal symptoms when alcohol consumption is stopped. Chronic alcohol abuse and dependence affects the whole body, and might elicit liver or lung diseases, might compromise immune and endocrine functions, and affects the brain. Alcoholism is a complicated disorder that is often characterized with refractoriness to treatment, chronic course, high comorbidity (Kozaric-Kovacic et al., 1992; Kozaric-Kovacic et al., 1995) and frequently with aggressive and suicidal behavior and suicide mortality. The lifetime mortality due to suicide in alcohol dependence ranges between 2% and 8%, and alcoholic patients have a 60–120 times greater suicide risk than the non-psychiatrically ill subjects (Sher, 2005). The comorbid depression in subjects with a history of alcoholism increases the suicidal risk. Since there is a large population of alcoholics in modern societies, the risk of suicides increases. Suicidal behaviors have the devastating effects on individuals, families, and society (Sher, 2005).

The neurobiological research in alcoholism may be divided into baseline studies, pharmacological challenge studies, which asses the biological responses after a challenge, and post-mortem studies (Ratsma et al., 2002). All these studies have their advantages and disadvantages, and these studies might help to discover individual variations in the trait markers, or toxic effects of alcohol on the neurobiological systems. The neurobiological basis of alcoholism is still not fully understood, but it is assumed to be associated with altered functions of the different neurotransmitter systems: opioid (ß-endorphin), gamma-amino butyric acid (GABA), dopamine or serotonin (5-hydroxytryptamine, 5-HT) systems (Lovinger, 1997; Johnson and Ait-Daoud, 2000; Ratsma et al., 2002). It has been hypothesized that there are four independent, genetically influenced intermediate phenotypes that affect the risk for alcoholism, and these are alcohol-metabolizing enzymes, trait markers like disinhibition and/or impulsivity, several independent psychiatric disorders (bipolar disorder), and the intensity of response to alcohol (Schuckit et al., 2004).

Scheme 1. Alcoholism.

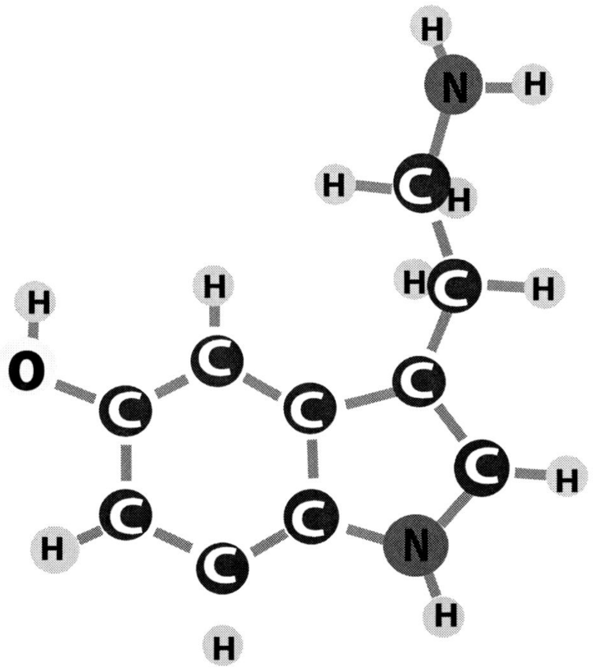

Scheme 2. Structural form of 5-HT.

Alcoholism is a complex disease, and multiple genetic but also environmental factors contribute to alcohol dependence (Goldman, 1995) Studies using twins or families with a high incidence of alcoholism, or children of alcoholics have identified specific regions of chromosomes associated with an increased risk of developing alcohol dependence, or individual genes or candidate genes. In this research the goal is a characterization of the specific endophenotypes, which are links between the manifestation of dependence and its biological underpinnings, and an identifying the specific regions of chromosomes associated with an increased risk of developing alcohol dependence, or individual genes or candidate genes within these regions. The functional polymorphisms in the enzymes that metabolize alcohol, such as alcohol dehydrogenase and mitochondrial aldehyde dehydrogenase are assumed to protect specific populations. The specific genes associated with the increased risk of alcoholism, relapse and addictive behavior are GABA receptor subunits GABRA2 and GABRG3, the muscarinic cholinergic receptor, and some candidate genes are specific alleles of dopamine, catechol-O-methyltransferase (COMT), neuropeptide Y, m-opioid receptor (OPRM), and serotonin transporter HTTLPR (Schuckit et al., 2004; Goldman et al., 2005;

Volkow and Li, 2005; Edenberg and Kranzler, 2005). Therefore, the search and the identification of the potential markers, or contributors to risk, such as individual biochemical and genetic characteristics, that might serve as state or trait markers for assessing the current level of alcohol consumption, or the predisposition to alcohol dependence of the subjects, is very important in alcohol research. These biomarkers might be used in the preclinical screening, diagnosis, disease staging, and predicting and monitoring of treatment. In alcoholism it is a very important tool to find the state markers of heavy alcohol consumption, since they can be used to detect early the harmful alcohol consumption, and to promote prevention and intervention. According to Ratsma et al. (2002) «a trait or vulnerability marker may be defined as a 'heritable trait, associated with a causative pathophysiological factor in an inherited disease'». The difference between a trait marker and a diagnostic marker is that trait marker does not need to have a high sensitivity or high specificity to a certain state or clinical entity (Ratsma et al., 2002). A trait marker has to be heritable, related to disease in the general population, and state independent (Ratsma et al., 2002).

1.2. Serotonin

1.2.1. Serotonin

Serotonin (5-hydroxytryptamine, 5-HT) is a heterocyclic amine (Scheme 2). Erspamer and Vialli in 1937 isolated a substance from the enterocromaffine mucosa of the gastrointestinal tract, with the ability to cause smooth muscle contraction (Erspamer and Asero, 1952). Later that heterocyclic amine was named 5-HT, and its many physiological roles were discovered. It is widespread in nature among multicellular eukaryotes, including plants (fruits, vegetables and nuts) and animals (both invertebrates and vertebrates). The main stores of 5-HT are enterochromaffine cells from the gastrointestinal mucosa, 5-HT neurons in the central nervous system, pineal gland and platelets (Kema et al., 2000).

Biosynthesis of 5-HT in human body takes place in the serotonergic neurons in brain, spinal cord, enterocromaffine cells of the gastrointestinal tract and in pineal gland, bronchi, thyroid, pancreas and thymus, and also in some neoplastic tissues (carcinoid tumors and phaeocromocytoma). There are two enzymes that are involved in 5-HT synthetic pathway, namely tryptophan hydroxylase and aromatic-l-amino acid decarboxylase, which convert amino acid tryptophan to 5-

HT (Scheme 3). Rate limiting steps of the biosynthesis are availability of tryptophan and 5-hydroxytryptophan (5-HPTP) formation. After biosynthesis 5-HT is stored mainly in the serotonergic neurons, gastrointestinal enterocromaffine cells, pineal gland and platelets. Catabolism of 5-HT is conducted in several ways. Enzyme monoamine oxidase (MAO) performs oxidative deamination of the lateral amino acid chain leading to formation 5-hydroxy-indol-acetylaldehyde, which is subsequently oxidized into 5-hydroxy-indol-acetic acid (5-HIAA), or reduced into 5-hydroxytriptophol. Minor pathway is conjugation with sulfuric or glucuronic acid. Also 5-HT can be N-acetylated and subsequently 5-O-methylated to form melatonin mainly in pineal gland (Scheme 3)

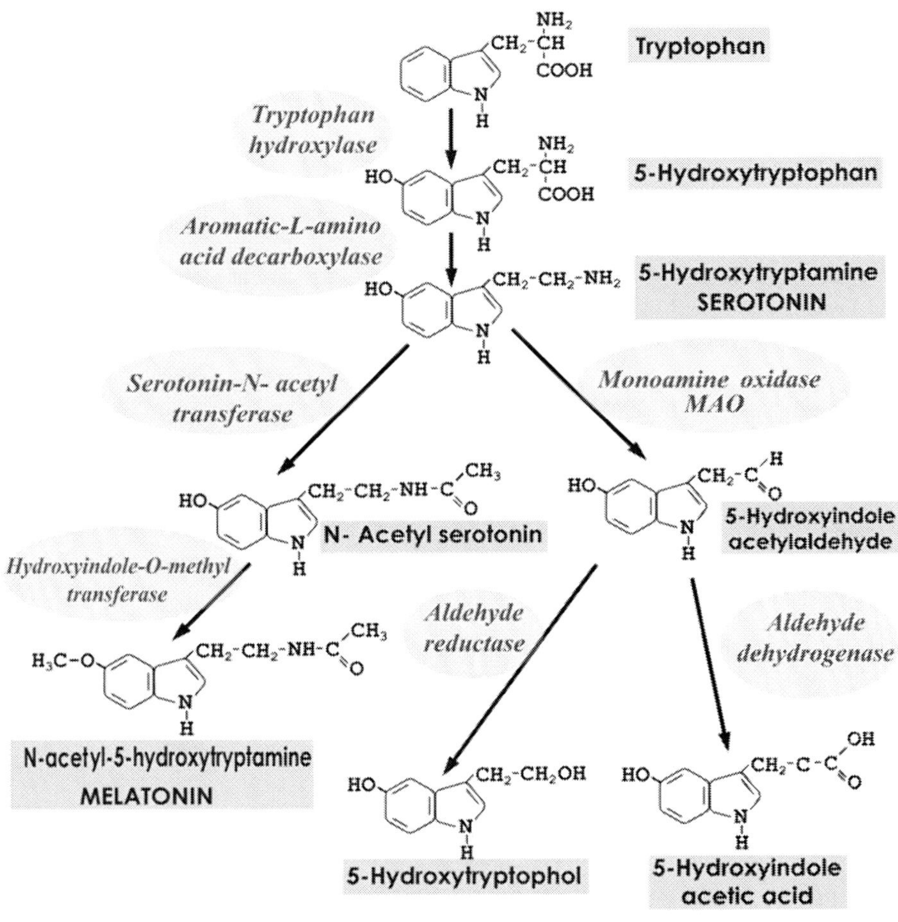

Scheme 3. Metabolism of 5-HT.

1.2.2. Physiology and Pathophysiology

Serotonin is synthesized in both central nervous system and periphery. Since it can not cross blood-brain barrier, there are two distinct 5-HT compartments. In central nervous system 5-HT functions as a neurotransmitter, neuromodulator, neurotrophic and neuroprotective agent. It regulates important behaviors such as feeding, sexual functions, hallucinatory behavior, sleep pattern, mood, emotions and sensory pathways including nociception (Lucki, 1998; Stahl, 1998). Serotonin has an important role in the processes of learning and memory, i.e. in the cognitive functions (Mattson et al., 2004).

During brain development 5-HT regulates synaptogenesis, neurite outgrowth and cell survival. In addition, in adult brain 5-HT regulates synaptic plasticity, neurogenesis and neuronal survival partly due to the ability to stimulate brain derived neurotrophic factor (BDNF) expression.

Serotonergic neurons in the central nervous system originate from the brainstem raphe nuclei (dorsal and median raphe nucleus), and project diffusely to limbic structures and neocortex. Neurons from dorsal raphe innervate amygdala, nucleus accumbens and other forebrain regions, and mediate anxiety and emotional reactions (Scheme 4). The projections from median raphe into forebrain innervate hippocampus and mediate tolerance to aversive stimuli and exert relaxation. The Scheme 4 shows 5-HT neurons: the cell bodies are in nuclei raphe (dorsal raphe and the raphe magnus), and its diverse projections.

In addition, 5-HT has an important neuroendocrine role, and 5-HT stimulates the stress system. Serotonin regulates the activity of the hypothalamic-pituitary-adrenal (HPA) axis (Dinan, 1996). The 5-HT projections from both dorsal and median raphe, together with serotonergic neurons from hypothalamus, innervate corticotrophin releasing hormone/factor (CRH/ CRH) neurons in the paraventricular nucleus of hypothalamus. Serotonin regulates the stress HPA axis system by the actions achieved through higher brain centers, paraventricular nucleus of the hypothalamus, and directly through pituitary. Its stimulates the CRH neurons and these actions trigger the release of CRH, which lead to activation of HPA axis with consequent release of adrenocorticotrophic hormone (ACTH) and glucocorticoid secretion. Thus 5-HT positively controls the HPA axis activity, and on the other hand, glucocorticoids and catecholamines mediate changes of the central 5-HT system induced by stress (Dinan, 1996).

Impairment of the 5-HT neurotransmission in the central nervous system elicits a variety of behavioral and physiological abnormalities and psychiatric disorders, including depression, autism, eating disorders, schizophrenia, obsessive/compulsive disorder, premenstrual syndrome, anxiety, panic disorder, seasonal affective disorder, extreme violence, hostility and aggression, suicide, migraine, manic depression, addiction, depressed mood, impulsivity, and more (Lucki, 1998; Stahl, 1998). The degeneration of the 5-HT neurons has been detected in Alzheimer's and Parkinson's diseases.

In periphery 5-HT functions as a vasoconstrictor and a proaggregator of platelets, neurotransmitter in the gut enteric plexus, and as an autocrine hormone of gastrointestinal enterocromaffine tract. The major source of peripheral 5-HT (80%) is enterocromaffine cells of the gastrointestinal tract, and the rest is absorbed from 5-HT rich food. Part of 5-HT enters circulation where it is mainly taken up by platelets, a portion is catabolized by liver or taken up by pulmonary endothelial cells where it is subsequently degradated, but also minor part stays in circulation conjugated to specific macromolecules. Peripheral 5-HT pathology includes hypertension, cardiac arrhythmia, drug-induced emesis and other disorders (Kema et al., 2000). Serotonin has a role in smooth muscle contraction and blood pressure regulation.

1.2.3. Serotonin Binding Sites, Receptors and Serotonin Transporter

Serotonin mediates its divergent functions through the binding to 5-HT receptors which are widely distributed throughout central and peripheral nervous systems, gastrointestinal tract, cardiovascular system and blood. There are seven distinct classes of serotonin receptors (5-HT1-5-HT7), shown in Tables 1 and 2. Almost all of them are G-protein coupled receptors with an exception of 5-HT3 class which is a ligand gated ion channel. Part of G-protein coupled receptors achieve their effector function via functional coupling to adenyl cyclase by Gi (inhibitory G protein) or Gs (stimulatory G protein) protein. Inhibitory G protein inhibits adenyl cyclase thus inhibiting cAMP formation and also acts to open K^+ channels in plasma membrane. On the other hand, Gs protein activates adenyl cyclase to produce cAMP and to open Ca^{2+} channels. Some G-protein linked 5-HT receptors activate inositide-specific phospholipase C via coupling by Gq protein. This enzyme cleaves phosphatidyl-inositol biphosphate in plasma membrane and form inositol triphosphate (IP3) and diacylglycerol. All these molecules (cAMP, IP3, diacylglycerol) serve as second messengers in intracellular signaling cascade acting upon target proteins thus influencing mechanisms controlling cell's transcription and translation, due to extracellular signal.

The binding of 5-HT to 5-HT3 receptor causes influx of Na+ that depolarize postsynaptic membrane, thus producing local permeability changes toward threshold polarity for initialization of action potential.

Serotonin transporter (5-HTT) is located in the serotonergic axon terminals, and in platelet membranes, with the main function to terminate the effects of 5-HT in the synapse. In platelet membrane 5HTT takes up 5-HT from the blood.

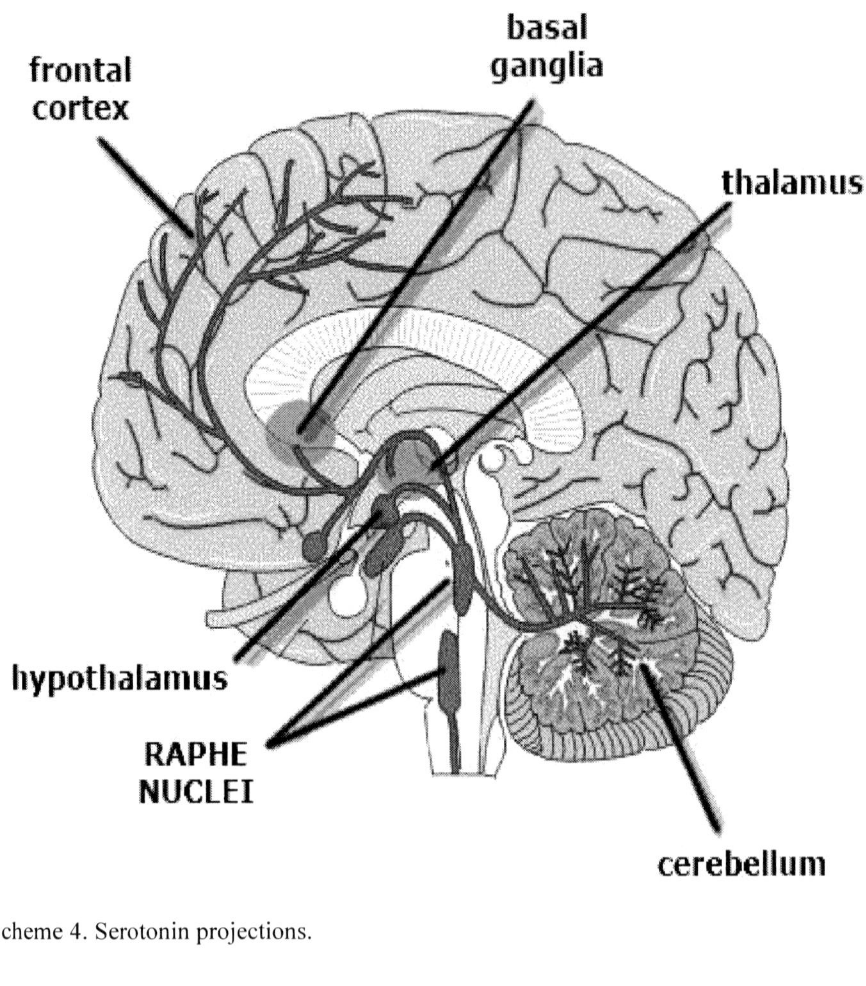

Scheme 4. Serotonin projections.

1.2.4. Serotonin Receptors and Functions: Serotonin Binding Sites - Receptors and Serotonin Transporter

Table 1. 5-HT1 and 5-HT2 receptor subtypes

Receptor	Effector mechanism		Human gene locus	Tissue and cell distribution		Mode of action	Clinical significance
5 - H T 1	Gi- protein coupled receptors	A	5q11.2-q13	raphe nuclei	somatodendritic autoreceptors	regulation of ACTH secretion, anxiety behaviors, temperature regulation	Depression, OCD, panic disorder, social phobia, eating disorders
				limbic structures: hypothalamus, amygdala, hippocampus, neocortical regions	Postsynaptic receptors		
		B	6q13	CNS: basal ganglia, striatum, frontal cortex, vascular tissues	terminal autoreceptors		
					heteroreceptors	control of release of other neurotransmitters (Ach, Glu, DA, NA, GABA)	
		D	1p34.3-p36.3	dorsal raphe nuclei, heart	autoreceptors	headache	migraine
		E	6q14-q15	CNS	?	?	
		F	3p11	raphe nuclei, hippocampus, striatum, thalamus, hypothalamus, neocortical regions, mesentery and uterus	autoreceptors ?		

Table 1. 5-HT1 and 5-HT2 receptor subtypes

Receptor	Effector mechanism		Human gene locus		Tissue and cell distribution		Mode of action	Clinical significance
5-HT2	Gq- protein coupled receptors	A	13q14-q21	basal ganglia, neocortical regions, platelet, claustrum, peripheral tissues	postsynaptic receptors	stimulation of hormone secretion (ACTH, corticosterone, oxytocin, renin, prolactin), platelet aggregation, hallucinations (LSD)	anxiety behaviors, temperature regulation, sexual functioning, sleep, psychosis, panic,	Schizophrenia - agitation, akathisia, anxiety, panic attacks, insomnia, sexual dysfunction, OCD
		B	2q36.3-q37.1	cerebellum, lateral septum, hypothalamus, medial amygdale, respiratory system, blood vessels		Vasorelaxation		
		C	Xq24	CNS, chorioid plexus		Hallucinations (LSD)		

ACTH-adrenocorticotrophic hormone; Ach- acetylcholine; CNS- central nervous system; DA- dopamine; GABA- γ-aminobutyric acid; Glu- glutamate; GI- gastrointestinal; NA- noradrenalin; OCD- obsessive compulsive disorder.

Table 2. 5-HT (5-HT$_3$ - 5-HT$_7$) receptor subtypes

Receptor	Effector mechanism		Human gene locus	Tissue and cell distribution		Mode of action	Clinical significances
5-HT3	ligand gated channel		11q23.1-q23.2	hippocampus, area posterma, dorsal motor nucleus of solitary tract, autonomic neurons, sensory system neurons	postsynaptic receptors	nausea, appetite, GI motility, vomiting	nausea, diarrhea, GI distress, headache
5-HT4	Gs- protein coupled receptors		5q31-q33	CNS, GI tract, heart	postsynaptic receptors	memory ?, Ach release in gut, heart rate	
5-HT5	?	A	7q36.1	astrocytes: hypothalamus, hippocampus, corpus callosum, fimbria, cerebral ventricles, glia	?	adaptive behavior under stressful situations	
5-HT6	Gs- protein coupled receptors		1p35-p36	striatum, amygdala, nucleus accumbens, hippocampus, neocortical regions, olfactory tuberculuc	postsynaptic receptors	control of central Ach ?, feeding ?	
5-HT7	Gs- protein coupled receptors		10q23.3-q24.4	CNS, blood vessels, smooth muscle	postsynaptic receptors	sleep	affective disorders

ACTH-adrenocorticotrophic hormone; Ach- acetylcholine; CNS- central nervous system; DA- dopamine; GABA- γ-aminobutyric acid; Glu- glutamate; GI- gastrointestinal; NA- noradrenalin; OCD- obsessive compulsive disorder.

1.2.5. Blood Platelets and 5-HT

The idea of using blood platelets as a potential model of serotonergic neurons has emerged after the discovery that psychoactive drugs such as reserpine deplete 5-HT, while imipramine inhibits the uptake of 5-HT in brain and platelets (Hrdina, 1994). These discoveries, together with the fact that measuring of the dysfunctions in the central nervous system are limited to the few (expensive) methods, such as functional imaging techniques: single photon emission computerized tomography (SPECT), positron emission tomography (PET), functional magnetic resonance imaging (FMRI), macroscopic neuropathology (computed tomography (CT), that may help in studying the anatomical and functional changes that characterize the process of drug addiction in the human brain, have elicited a lot of research with the assumption that blood platelets might be used as limited peripheral model for the central 5-HT neurons (Stahl, 1985; Andres et al., 1993; Hrdina, 1994; Camacho and Dimsdale, 2000; Mendelson, 2000; Goveas et al., 2004). Blood platelet share some similarities, and differences, with the central 5-HT synaptosomes (Scheme 5).

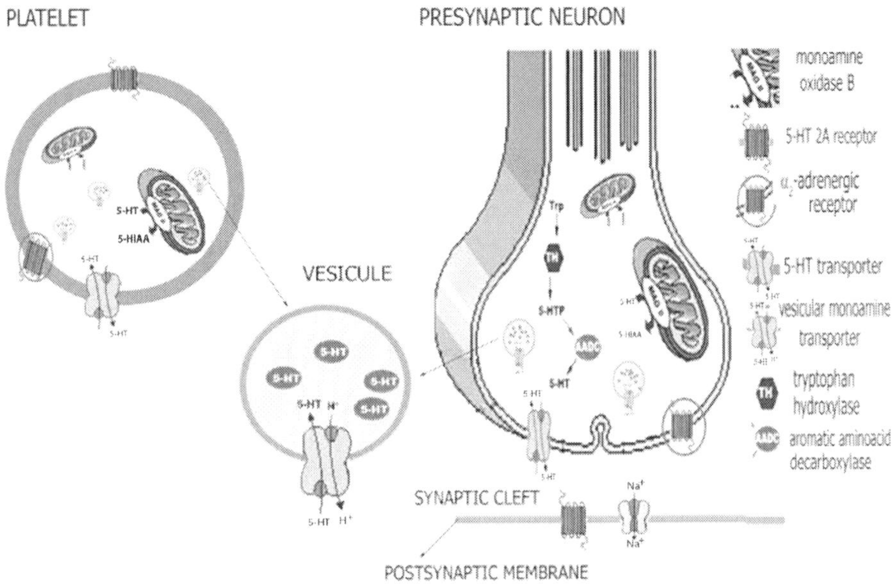

Scheme 5. Similarities and differences between blood platelets and serotonergic synaptosomes.

Blood platelets have the similar pharmacodynamics of 5-HT with central 5-HT neurons (Mueller-Oerlinghausen et al., 2004). The similarities between neurons and platelets (Scheme 5) are in uptake, storage and release of 5-HT, platelet monoamine oxidase (MAO) type B, 5-HT transporters (Lesch et al., 1994), 5-HT2 receptors (Andres et al., 1993), α_2-adrenergic receptors, and binding sites for ^{3}H-imipramine, ^{3}H-paroxetine, and ^{3}H-LSD, and signaling mechanism which share identical pharmacologic and kinetic characteristics with the central nervous receptors and binding sites, and resemble the corresponding counterparts and biological processes in the central serotonergic neurons. Platelets differ from the 5-HT neurons since platelets origin from mesoderm and neurons from ectoderm, platelets lack nucleus, and they do not biosynthesize 5-HT, and in the function (Hrdina, 1994).

1.2.6. Platelet MAO

Monoamine oxidase is an enzyme that has two isoenzymes: type A and type B. MAO-A and MAO-B are both target enzymes for the deamination of different neurotransmitter and xenobiotic amines in the central nervous system and in the periphery. MAO-B type is located in neurons and platelets, and this isoenzyme from the platelets (MAO-B or platelet MAO) has the identical amino acid sequence with MAO-B from the central nervous system (Billet, 2004; Oreland 2004). Platelet MAO shares similar biochemical and pharmacological characteristics with brain MAO-B (Coccini et al., 2005). MAO-B is a flavoenzyme that oxidatively deaminates exogenous monoamines such as phenyletylamine and benzylamine, and dopamine and tyramine are substrates for both MAO types. Platelet MAO has been proposed to represent a biological marker of the different psychopathology and personality (Oreland and Shasken, 1983; Oreland, 2004), and a genetic marker for the size or functional capacity of the central serotonergic system (Oreland, 2004). Platelet MAO activity is reported to be associated with personality traits such as impulsiveness, extroversion, sensation seeking, neuroticism, or novelty seeking (Kozaric-Kovacic et al., 2000; Oreland et al., 2002; Oreland, 2004), impulse and affect dysregulation (Verkes et al., 1998), and might represent a biological marker for the personality traits that increase the vulnerability for drug-abuse, social mal-adaptation or dysinhibitory psychopathology (Oreland et al., 2002; Oreland, 2004; Longato-Stadler et al., 2002). Decreased MAO-B activity was reported to be the characteristic feature of impulsiveness, sensation-seeking and aggressiveness (Buchsbaum et al., 1976; Schalling et al., 1987), or to various psychopathological conditions, suicidal behavior, hyperactivity, and type II alcoholism (Schalling et al., 1987; von Knorring et al., 1991; Devor et al., 1993; Oreland and Hallman, 1995). There are still a lot of controversies regarding MAO-B, since MAO-B is expressed in a cell- and tissue-specific manner (Billett, 2004; Costa-Mallen et al., 2005), and no association was found in the activity of MAO-B in platelets and the brain (Winblad et al., 1979). This may indicate that MAO-B might be regulated differently in the blood and brain (Ekblom et al., 1998).

1.2.7. Alcoholism and Serotonin

The dysfunction in 5-HT neurotransmission may be associated with a propensity toward alcohol abuse and dependence (Johnson, 2004). The disturbance in central 5-HT system in alcoholism might result from the dysfunction of serotonergic transmission, which might be a predisposing factor to excessive alcohol consumption and alcohol dependence. A reduction in brain-stem 5-HT transporters in abstinent alcoholics when compared to control subjects was found using SPECT (Heinz et al., 1998). Low activity of the central 5-HT neurotransmission has been associated with alcohol drinking behavior (LeMarquand et al., 1994). The role of 5-HT in alcoholism is based on the data showing alterations in the central and peripheral indices of 5-HT function. The uptake of 5-HT is altered in alcohol dependent subjects (Heinz et al., 2001; Kent et al., 1985). The 5-HT uptake has been reported to be lower in alcoholics and former alcoholics than in control subjects (Kent et al., 1985), and increased (Faraj et al., 1997), or unaltered in alcoholics when compared to healthy controls (Javors et al., 2000). The discrepancies might be due to the early-onset and late-onset of alcoholism, since those with early-onset show higher platelet 5-HT transporter activity than those with late-onset of alcoholism (Javors et al., 2000). Alcoholism-induced fall of 5-HT transport has been related to decreased platelet 5-HT content.

Neurochemistry of alcoholism was investigated also using post-mortem brain tissue from alcoholics. A reduced 5-HT transporter binding in the hippocampus (Gross-Isseroff and Biegon, 1988; Chen et al., 1991), and reduced binding in the 5-HT1A receptors of alcoholics compared to control subjects (Dillon et al., 1991) was found. A postmortem whole-hemisphere autoradiography showed lower 5-HT transporter binding in the caudate body of alcoholics compared to nonalcoholic control subjects, indicating that the 5-HT transporter binding in the dorsal striatum may be lower in alcoholics (Storvik et al., 2006).

Challenge studies have shown that hormonal probes used to assess central 5-HT function show deficits in the central 5-HT functions in alcoholic subjects: prolactin or cortisol response to administration of 5-HT drugs fenfluramine (Balldin et al., 1994), m-clorophenylpiperazine (Krystal et al., 1996), and 6-chloro-2-1-piperazinylpyrazine (Lee and Meltzer, 1991) was blunted, adrenocorticotropic hormone (ACTH) response to m-chlorophenylpiperazine (mCPP) was lower in alcoholic than in nonalcoholic patients (Krystal et al., 1994).

Lower levels of CSF 5-HIAA levels are proposed to indicate a central 5-HT disturbance in the early-onset alcoholics (Linnoila et al., 1983). These data are corroborated with the findings showing lower 5-HIAA CSF levels in abstinent alcoholics (Ballenger et al., 1979), which returned to normal values after 7 days of drinking (Zarcone et al., 1975), and reduced CSF 5-HIAA in abstinent alcoholics (Banki, 1981), in alcoholic impulsive offenders (Virkkunen et al., 1995), and in early-onset alcoholics (Fils-Aime et al., 1996). In line with that, low precursor 5-HT availability was found in early-onset alcoholism (Buydens-Branchey et al., 1989). Alcoholic subjects have decreased plasma tryptophan levels (Branchey et al., 1981).

In addition, 5-HT regulates divergent behaviors (Stahl, 1998; Lucki 1998), disturbed in personality disorders, addiction and alcoholism (Coccaro et al., 1996; Twitchell et al., 1998; Moeller and Dougherty, 2001; Schmidt et al., 1997; Soloff et al., 2000), and has an important role in suicidal behavior (Roy and Linoilla, 1988; Mann et al., 1996). Suicide is a major complication of the psychiatric disorders, and a major cause of death throughout the world. According to the WHO data, suicide accounts for almost 2% of the deaths in the world (Bondy et al., 2006). Suicidal behavior, often comorbid with alcoholism, also points to the involvement of serotonergic dysfunction in alcoholism (Dubovsky and Thomas, 1995; Gruenewald et al., 1995; Veenstra-Vader Weele et al., 2000; Gorwood, 2001; Muller-Oerlinghausen et al., 2004). Although the causal relationship between 5-HT and suicidal behavior has been suggested (Muck-Seler et al., 1996a; Pivac et al., 1997a) but also questioned (Mueller-Oerlinghausen et al., 2004), indices of 5-HT function have been altered in suicidal behavior. Suicidality is related to a decreased central serotonergic function (Mann, 1998), with reduced cerebrospinal fluid 5-HIAA concentrations in depressed suicidal patients, especially in violent suicide attempters, compared to depressed non-suicidal patients (Asberg et al., 1976; Mann and Malone, 1997), and reduced levels of 5-HT and 5-HIAA in blood and CSF of suicide attempters were found (Ballenger et al. 1979). Low CSF 5-HIAA has been found in intoxicated violent offenders (Virkkunen et al., 1995). Suicidality is associated with a deficient 5-HT system (Mann et al., 1996; Cremniter et al., 1994; Muck-Seler et al., 1996a; Pivac et al., 1997a).

Besides in alcoholism and suicidal behavior, 5-HT has an important role in antisocial and aggressive behavior (Popova, 2006). Subjects with antisocial and aggressive behavior have lower 5-HT levels than nonaggressive individuals (Coccaro et al., 1989). In the studies using 5-HT challenge tests, a 5-HT1A agonist buspirone elicited a blunted prolactin response in violent compared to nonviolent parolees (Cherek et al., 1999). The lowered 5-HIAA in antisocial when compared to non-antisocial adults was confirmed in a recent meta-analysis (Moore et al., 2002).

Blood platelets might be used as a limited peripheral model for the central 5-HT synaptosomes, since they share similar biochemical processes (Stahl, 1985; Andres et al., 1993; Camacho and Dimsdale, 2000; Muller-Oerlinghausen et al., 2004). Platelet 5-HT concentration has been shown to be decreased in suicidal compared to non-suicidal patients or healthy controls (Alvarez et al., 1999), and lower levels of whole blood 5-HT have been related to suicidality in patients with schizophrenia (Braunig et al., 1989), schizoaffective disorder (Rao et al., 1994), or depression (Muck-Seler et al., 1996a; Pivac et al., 1997a). On the other hand, platelet 5-HT concentration has been found to be increased in psychotic depression (Muck-Seler et al., 1996a; Pivac et al., 1997a), psychotic mania (Sagud et al., 2006), PTSD with psychotic features (Pivac et al., 2006a), or positive symptoms in schizophrenia (Jakovljevic et al., 1997; Pivac et al., 1997b).

Peripheral indices of 5-HT function have been reported to be altered in alcoholism. The majority of studies evaluated platelet serotonergic measures in acutely intoxicated alcoholics (Schmidt et al., 1997) or in alcoholics during withdrawal (Bailly et al., 1990; Bailly et al., 1993; Pivac et al., 2004). Platelet serotonergic markers, such as platelet 5-HT or platelet MAO, have been associated to particular trait markers of psychopathology (Askenazy et al., 2000; Mueller-Oerlinghausen et al., 2004). Platelet 5-HT concentration (Oreland, 2004), and platelet MAO activity (Schalling et al., 1987; Garpenstrand et al., 2000; Oreland, 2004; Kozaric-Kovacic et al., 2000), have been proposed to serve as the trait markers for particular mental disturbances. Serotonergic alterations might contribute to the cognitive disturbances and deficits in the memory systems occurring in alcoholism, and platelet 5-HT has been reported to be altered in aggression (Goveas et al., 2004), impulsivity (Askenazy et al., 2000), and suicidal behavior (Muck-Seler et al., 1996a, Pivac et al., 1997a), behaviors closely related to alcoholism. Platelet MAO has been proposed to be a biological marker of the psychopathology and personality (Oreland and Shasken, 1983; Oreland, 2004). Studies of the possible association between alcoholism and platelet MAO activity have been inconsistent, with decreased (Oreland, 2004) or similar (Farren et al., 1998; Pivac et al., 2005) platelet MAO values found in alcoholic subjects when compared to healthy controls. These discrepancies might be due to the influence of different factors that affect platelet MAO activity, such as sex (Muck-Seler et al., 1991; Veral et al., 1997; Pivac et al., 2005), age (Veral et al., 1997), smoking status (Oreland et al., 2002; Fowler et al., 2003), lithium (Oreland, 2004), or neuroleptic (Verkes et al., 1998) treatment.

Since the role of platelet serotonergic markers, such as platelet 5-HT and platelet MAO activity, as biological markers of alcoholism were proposed (Pivac et al., 2004; Oreland, 2004), and disputed (Pivac et al., 2005; Anthenelli et al., 1998), in the present study we compared platelet 5-HT concentration and platelet MAO activity in the large groups of drug naïve male and female alcoholic subjects, subdivided according to the different psychiatric comorbidities (personality disorder (PD), posttraumatic stress disorder (PTSD), anxious-depressive disorder (ADD), major depression, schizophrenia), and smoking status, with the corresponding male and female healthy subjects and male and female patients with personality disorder, PTSD, ADD, major depression, schizophrenia. The aims of the present study were: a) to determine platelet 5-HT concentration and platelet MAO activity in the large groups of drug-free alcoholics who were admitted into hospital due to the withdrawal syndrome related to alcohol dependency, and compare these values in the sex-matched healthy controls, b) to determine whether sex, smoking status, number of cigarettes smoked per day, or different psychiatric comorbid diagnoses (personality disorder, PTSD, ADD, major depression, schizophrenia), would affect platelet 5-HT concentration or platelet MAO in male and female alcoholic patients.

Chapter 2

Methods

2.1. Subjects

The diagnosis of alcohol dependency in 178 male and 42 female ethnically homogenous drug-free alcoholics was made by experienced psychiatrists in the Psychiatric Hospital Vrapče, Center for Alcoholism and Other Addictions. Subjects were evaluated by structured clinical interview, conducted by two psychiatrists, and final diagnoses were made according to the DSM-IV criteria (American Psychiatric Association, 1994). Patients were admitted into Centre for Alcoholism and Other Addictions in Psychiatric Hospital Vrapce, due to the withdrawal syndrome related to alcohol dependency. Withdrawal symptoms were of non-psychotic type, intensified 1 week before the admission. Most patients were acutely intoxicated with alcohol. Therefore, the interview and the sampling were done when the patients were sober, i.e. 10 h after the admission at the department. The mean age in alcoholic male subjects was 49.1 ± 10.1 years (range 28-76), and in alcoholic female subjects was 51.8 ± 10. 4 years (range 35-73). Demographic data of the patients are presented in table 3. All patients fulfilled the CAGE questionnaire (Ewing, 1984), and were evaluated by the Hamilton Rating Scale for Depression (HAMD) and Anxiety (HAMA), (Hamilton, 1960).

Besides the alcohol dependency that was a primary diagnosis, alcoholic subjects were diagnosed also with different psychiatric comorbid diagnoses (personality disorder, PTSD, ADD, major depression, schizophrenia). These psychiatric disorders were diagnosed according to the DSM-IV criteria (American Psychiatric Association 1994). Patients were not treated with any psychotropic drugs before the samples were selected. All participants gave their written informed consent. The local Ethic committee approved this protocol.

Control subjects comprised of 155 male and 178 female healthy volunteers. Healthy control males had mean age 38.34 ± 13.04 years (range 21-63), and females had 41.14 ± 13.17 years (range 19-59). The control group consisted of drug-free healthy volunteers, mainly medical staff, with no personal or family history of substance abuse or dependence, no hepatic or renal disorders, no medical treatment, and alcohol-free for at least 24 h.

2.2. Determination of Platelet 5-HT Concentration

The sampling was done when the patients were sober, i.e. 10 h after the admission at the department. A forearm vein was cannulated for blood sampling at 08.00 a.m. Blood samples (8 ml) were drawn in a plastic syringe with 2 ml of acid citrate dextrose anticoagulant. Platelet rich plasma (PRP) was obtained after centrifugation of whole blood at 950g for 30 sec at room temperature and platelets were sedimented by further centrifugation of PRP at 10,000g for 5 min in a refrigerated centrifuge. The platelet pellet was washed with saline and centrifuged again. Sedimented platelets were frozen and stored at -20°C no longer than 7 days before being assayed for 5HT concentration. Platelet 5-HT concentration was determined by the spectrofluorimetric method, as previously described (Pivac et al., 2001). Briefly, platelets were destroyed by sonication (20 kHz, amplitude 8×10^{-3} mm for 30 sec). Specimens of standard, blank (water) and platelet sonicates were analyzed in duplicate. All samples were deproteinized with 1 ml of 10% $ZnSO_4$ and 0.5 ml of 1 N NaOH. For the preparation of fluorophore, 0.2 ml of L-cysteine (0.1%) and 1.2 ml of orthophthalaldehyde (0.05%) were added to deproteinized samples. The measurement of the 5-HT fluorescence was performed on a Varian Spectrophotofluorimeter Cary Eclipse, on an exciting wavelength of 345 nm and emitted wavelength of 485 nm. Platelet protein concentrations were measured by the method of Lowry et al. (1951).

2.3. Determination of Platelet MAO Activity

Platelet MAO-B activity was determined spectrofluorimetrically using kynuramine as a substrate, by a slight modification of the method of Krajl (1965), as previously described (Pivac et al., 2002; Muck-Seler et al., 2002). Briefly, 100 μl of standard (4-hydroxyquinoline, 4-HOQ), blank (water) and platelet sonicates were incubated with 100 μl of kynuramine (final concentration 73.6 μM) at 37°C for 1 hour. The reaction was stopped with 1 N NaOH. The measurement of 4-HOQ fluorescence, a product of kynuramine, was performed on Varian Spectrophotofluorimeter Cary Eclipse, with an exciting wavelength of 310 nm and emitted wavelength of 380 nm. Platelet protein levels were measured by the method of Lowry et al. (1951).

2.3. Data Analysis

Statistical evaluation of the results, expressed as means ± standard deviations (SD), was performed using one-way analysis of variance (ANOVA) followed by the Student Newman Keuls multiple comparison test. Two-way ANOVA was used to test the significance of the several variables: sex, smoking, diagnosis, or different comorbid diagnoses, on platelet MAO activity, and to test the interactions between them. The correlation between parameters was determined by a Pearson's coefficient of correlation. The significance was accepted when $p<0.05$. The statistical analysis of the data was conducted with Statistica, release 6, Statsoft (Tulsa, Oklahoma, USA), and with Sigmastat 2.0 (Jandell Scientific Corp. San Raphael, California, USA).

Chapter 3

Results

3.1. Platelet 5-HT Concentration in Alcoholic Patients and Healthy Controls

Platelet 5-HT concentration was determined in 178 male alcoholic patients and in 42 female alcoholic patients and these values were compared with values in 155 male healthy subjects and in 171 female healthy control subjects.

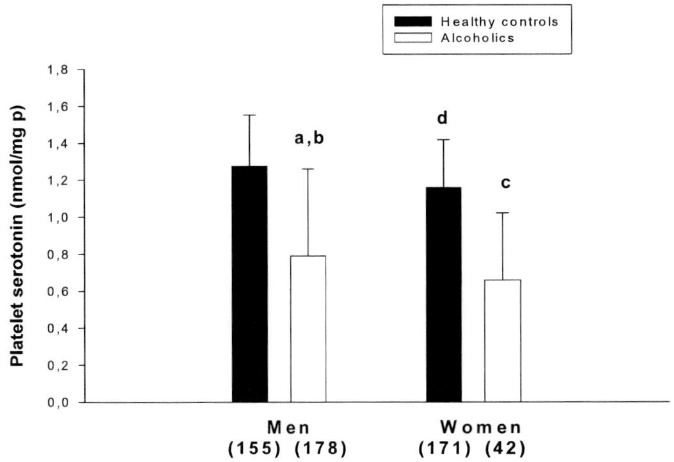

Figure 1. Platelet serotonin concentration in male and female alcoholic patients, and in male and female healthy controls. Each column represents mean ± SD. Number of subjects is given in brackets. [a] $p<0.001$ vs. healthy men; [b] $p<0.032$ vs. female alcoholics; [c] $p<0.001$ vs. healthy women; [d] $p<0.001$ vs. healthy men (Student Newman Keuls test).

Platelet 5-HT concentration differed significantly ($F=73.75$; $df=3,542$; $p<0.001$, ANOVA) between 178 male and 42 female alcoholic patients and 155 male and 171 female healthy controls (Figure 1). The between group comparisons (Newman Keuls test) showed that alcoholic male and female patients had significantly ($p<0.001$) lower platelet 5-HT concentration when compared to corresponding control men and women.

Sex differences were detected between healthy subjects and alcoholic patients, and female subjects had significantly lower platelet 5-HT concentration than male subjects. Platelet 5-HT concentration was also significantly ($p<0.032$) lower in female than in male alcoholic subjects. In addition, platelet 5-HT concentration was significantly ($p<0.003$) lower in female than in male healthy subjects (Figure 1).

3.2. Platelet 5-HT Concentration in Alcoholic Patients and Healthy Controls Divided According to the Smoking Status

Platelet 5-HT concentration did not differ significantly ($F=1.28$; $df=3,179$; $p=0.282$, ANOVA) between 142 male and 41 female alcoholic patients (Figure 2), who were divided, according to the smoking status, into smokers and nonsmokers. There were no significant differences in platelet 5-HT concentration between 98 male alcoholic smokers and 44 male alcoholic nonsmokers, or between 23 female alcoholic smokers and 18 female alcoholic nonsmokers. Therefore, the smoking status did not significantly alter platelet 5-HT concentration in male or female alcoholic subjects (Figure 2). Since smoking status did not significantly change platelet 5-HT concentration, alcoholic male or female patients were not divided into smokers and nonsmokers for further statistical evaluation of the data.

Platelet 5-HT concentration did not vary significantly ($F=1.09$; $df=3,153$; $p=0.356$, ANOVA) between 113 male and 44 female healthy subjects (Figure 3), who were divided based on their smoking status into smokers or nonsmokers. There were no significant differences in platelet 5-HT concentration between 44 male healthy smokers and 69 male healthy nonsmokers, or between 17 female healthy smokers and 27 female healthy nonsmokers. Therefore, the smoking status did not significantly change the values of platelet 5-HT in healthy men or women (Figure 3), and consequently healthy men or women were not divided into smokers and nonsmokers for further statistical evaluation of the data.

3.3. Platelet 5-HT Concentration in Alcoholic Patients Subdivided According to the Different Psychiatric Comorbidities, in Psychiatric Patients and in Healthy Controls

Platelet 5-HT concentration was significantly different ($F=39.80$; $df=5,500$; $p<0.001$, ANOVA) between 147 male healthy subjects, 115 male alcoholic patients without psychiatric comorbidities and 62 male alcoholics with psychiatric comorbidities, and between 140 female healthy controls, 16 female alcoholic patients without psychiatric comorbidities and 26 female alcoholics with psychiatric comorbidities (Figure 4).

When alcoholic patients were subdivided according to the different psychiatric comorbidities, their platelet 5-HT concentration was significantly lower ($p<0.001$) that the values in the corresponding healthy subjects.

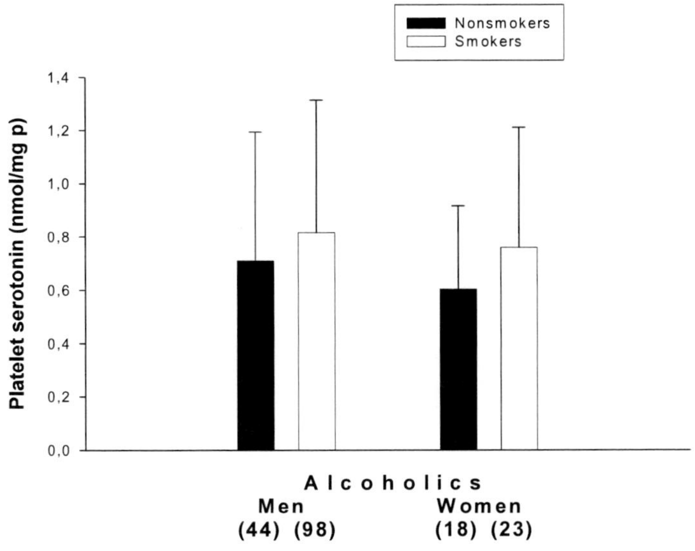

Figure 2. Platelet serotonin concentrations in male or female alcoholic subjects, who were smokers or nonsmokers. Each column represents mean ± SD. Number of subjects is given in brackets.

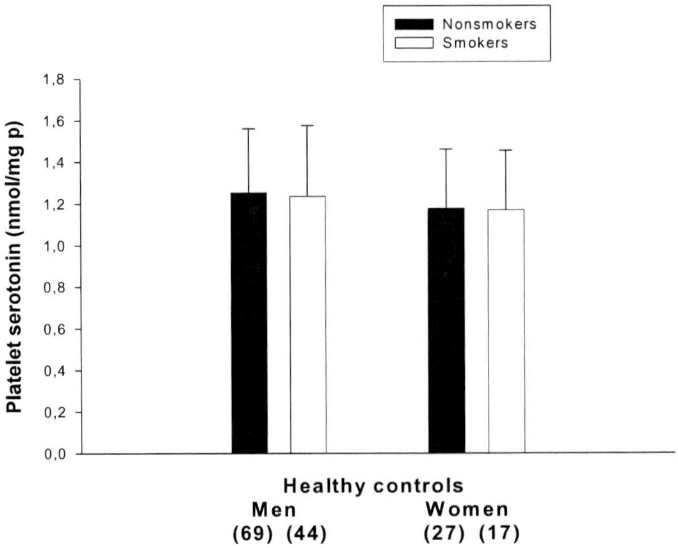

Figure 3. Platelet serotonin concentrations in male or female healthy subjects, who were smokers or nonsmokers. Each column represents mean ± SD. Number of subjects is given in brackets.

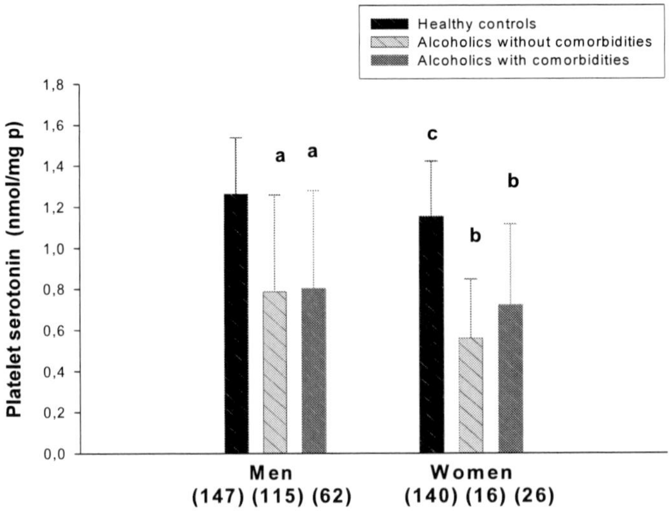

Figure 4. Platelet serotonin concentrations in male and female alcoholic patients with or without different psychiatric comorbidities, and in healthy male and female controls. Each column represents mean ± SD. Number of subjects is given in brackets.[a] p<0.001 vs. healthy men; [b] p<0.001 vs. healthy women; [c] p<0.01 vs. healthy men (Student Newman Keuls test).

Male alcoholic patients without psychiatric comorbidities and male alcoholics with psychiatric comorbidities had significantly lower ($p<0.001$) platelet 5-HT concentration than male healthy subjects. Female alcoholic patients without psychiatric comorbidities and female alcoholic patients with psychiatric comorbidities had significantly lower ($p<0.001$) platelet 5-HT concentration than female healthy subjects. Healthy men had significantly higher ($p<0.01$) platelet 5-HT concentration than healthy women (Figure 4).

3.4. Platelet 5-HT Concentration in Alcoholic Patients with PTSD, Alcoholic Patients without Psychiatric Comorbidities, Patients with PTSD and Healthy Controls

Platelet 5-HT concentration differed significantly ($F=32.21$; $df=3, 325$; $p<0.001$, ANOVA) between 147 male healthy subjects, 115 male alcoholic patients without psychiatric comorbidities, 19 male alcoholics with PTSD and 48 male patients with PTSD (Figure 5).

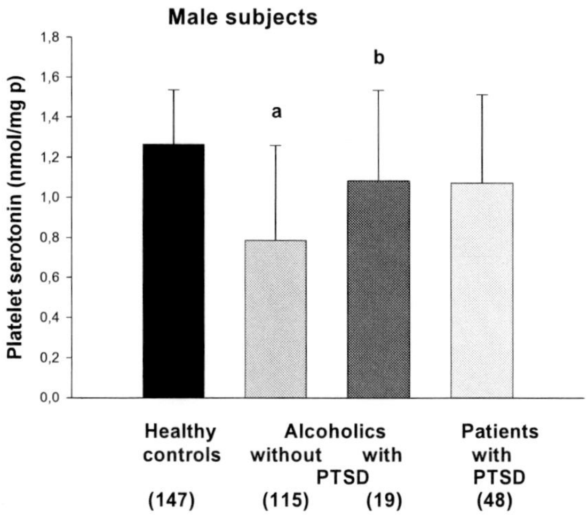

Figure 5. Platelet serotonin concentrations in male alcoholic patients with PTSD, alcoholic patients without psychiatric comorbidities, male patients with PTSD and male healthy controls. Each column represents mean ± SD. Number of subjects is given in brackets. [a] $p<0.001$ vs. healthy men; [b] $p<0.006$ vs. male alcoholic patients without PTSD (Student Newman Keuls test).

When male alcoholic patients were subdivided according to the comorbid PTSD, their platelet 5-HT concentration was not ($p=0.06$) significantly lower when compared to healthy men. However, alcoholic male patients with PTSD had significantly ($p<0.0016$) higher platelet 5-HT concentration than male alcoholic patients without psychiatric comorbidities. Platelet 5-HT concentration did not differ significantly ($p=0.91$) between male alcoholic patients with PTSD or PTSD patients (Figure 5).

3.5. Platelet 5-HT Concentration in Alcoholic Patients with Schizophrenia, Alcoholic Patients without Psychiatric Comorbidities, Schizophrenic Patients and Healthy Controls

Platelet 5-HT concentration differed significantly ($F=55.09$; $df=3, 248$; $p<0.001$, ANOVA) between 147 male healthy subjects, 115 male alcoholic patients without psychiatric comorbidities, 6 male alcoholics with schizophrenia, and 20 male patients with schizophrenia (Figure 6).

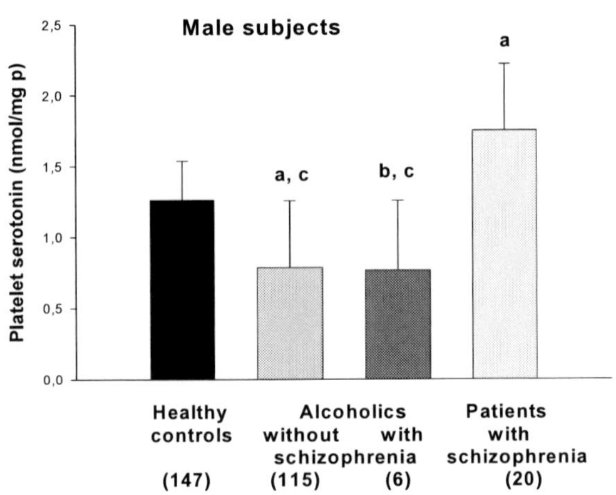

Figure 6. Platelet serotonin concentrations in male alcoholic patients with schizophrenia, in male alcoholic patients without any psychiatric comorbidities, in male schizophrenic patients and in male healthy controls. Each column represents mean ± SD. Number of subjects is given in brackets. [a] $p<0.001$ vs. healthy men; [b] $p<0.006$ vs. healthy men; [c] $p<0.001$ vs. male patients with schizophrenia (Student Newman Keuls test).

When male alcoholic patients were subdivided according to the comorbid schizophrenia, their platelet 5-HT concentration was similar ($p=0.91$) to the values in male alcoholic patients without psychiatric comorbidities, but was significantly ($p<0.006$) lower compared to values in healthy men. Alcoholic male patients without psychiatric comorbidities had significantly ($p<0.001$) lower platelet 5-HT concentration than healthy men. Male patients with schizophrenia had significantly higher ($p<0.01$) platelet 5-HT concentration than male healthy control subjects, or than male alcoholic patients with comorbid schizophrenia (Figure 6).

3.6. Platelet 5-HT Concentration in Alcoholic Patients with Personality Disorders, Alcoholic Patients without Psychiatric Comorbidities, Patients with Personality Disorders and Healthy Controls

Platelet 5-HT concentration differed significantly ($F=37.33$; $df=3, 287$; $p<0.001$, ANOVA) between 147 male healthy subjects, 115 male alcoholic patients without psychiatric comorbidities, 18 male alcoholics with personality disorders (PD), and 11 male patients with personality disorders (PD), (Figure 7).

When male alcoholic patients were subdivided according to the comorbid personality disorders (PD), they had similar ($p=0.74$) values of platelet 5-HT as male alcoholic patients without psychiatric comorbidities. Platelet 5-HT concentration in alcoholic patients with comorbid personality disorders (PD) was significantly ($p<0.001$) lower compared to values in healthy men, or to values in patients with personality disorders (PD). Alcoholic male patients without psychiatric comorbidities had significantly ($p<0.001$) lower platelet 5-HT concentration than healthy men (Figure 7).

Platelet 5-HT concentration differed significantly ($F=31.16$; $df=3, 170$; $p<0.001$, ANOVA) between 140 female healthy subjects, 16 female alcoholic patients without psychiatric comorbidities, 6 female alcoholic patients with personality disorders (PD), and 12 female patients with personality disorders (PD), (Figure 8).

Female alcoholic patients without psychiatric comorbidities and female alcoholic patients with personality disorders (PD) had significantly lower ($p<0.001$) platelet 5-HT concentration than female healthy subjects.

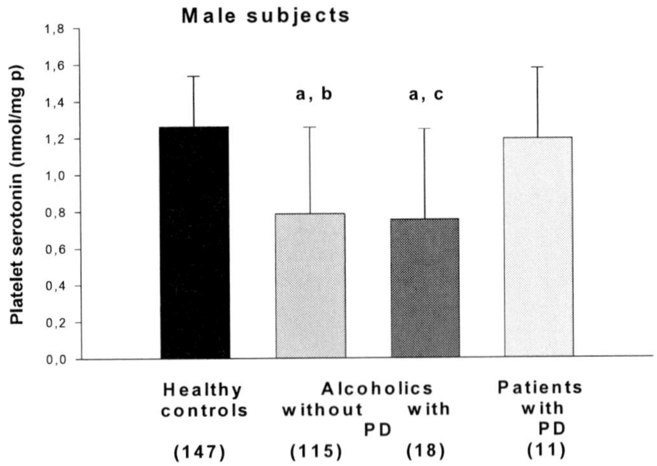

Figure 7. Platelet serotonin concentrations in male alcoholic patients with personality disorders (PD), in male alcoholic patients without any psychiatric comorbidities, in male patients with personality disorders (PD) and in male healthy controls. Each column represents mean ± SD. Number of subjects is given in brackets. [a] $p<0.001$ vs. healthy controls; [b] $p<0.001$vs. patients with personality disorders (PD); [c] $p=0.008$ vs. patients with personality disorders (PD); (Student Newman Keuls test).

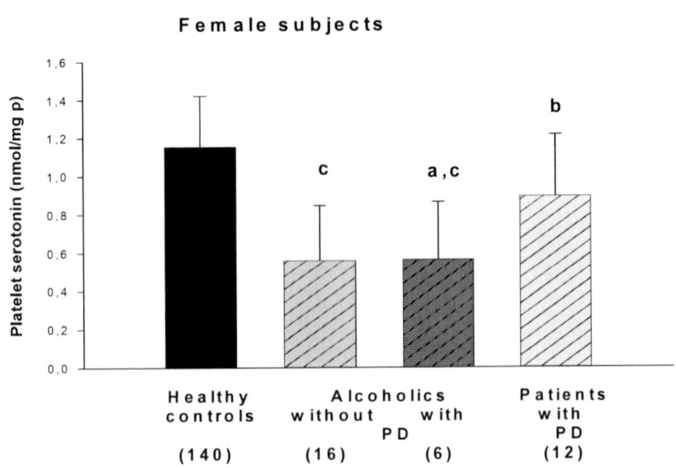

Figure 8. Platelet serotonin concentrations in female alcoholic patients with comorbid personality disorders (PD), female alcoholic patients without any psychiatric comorbidities, in female patients with personality disorders (PD) and in female healthy controls. Each column represents mean ± SD. Number of subjects is given in brackets. [a] $p=0.016$ vs. patients with personality disorder (PD); [b] $p=0.002$ vs. healthy controls; [c] $p<0.001$ vs. healthy controls (Student Newman Keuls test).

Female patients with personality disorders (PD) had significantly higher ($p<0.016$) platelet 5-HT concentration than female alcoholic patients with personality disorders (PD). Female alcoholic patients without psychiatric comorbidities and female alcoholic patients with personality disorders (PD) had similar ($p=0.97$) platelet 5-HT concentration (Figure 8).

3.7. Platelet 5-HT Concentration in Alcoholic Patients with Depression, Alcoholic Patients without Psychiatric Comorbidities, Depressed Patients and Healthy Controls

Platelet 5-HT concentration differed significantly ($F=37.37$; $df=3, 295$; $p<0.001$, ANOVA) between 147 male healthy subjects, 115 male alcoholic patients without psychiatric comorbidities, 3 male alcoholics with depression, and 34 male patients with depression (Figure 9).

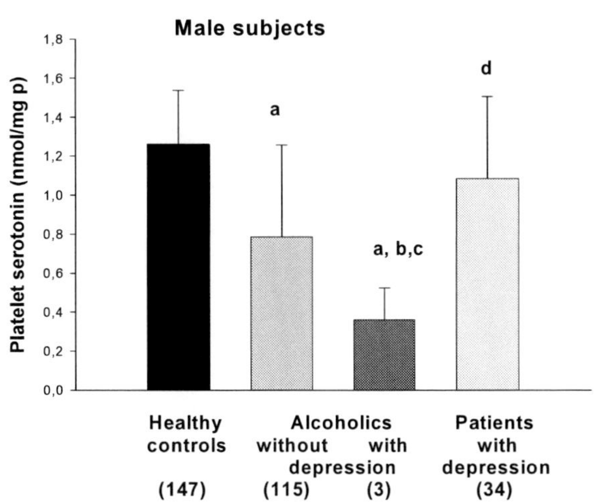

Figure 9. Platelet serotonin concentrations in male alcoholic patients with comorbid depression, male alcoholic patients without any psychiatric comorbidities, male patients with depression and healthy male controls. Each column represents mean ± SD. Number of subjects is given in brackets. [a] $p<0.014$ vs. healthy controls; [b] $p=0.004$ vs. alcoholics without depression; [c] $p=0.001$ vs. healthy controls; [d] $p<0.001$ vs. alcoholic patients with depression (Student Newman Keuls test).

Female alcoholic patients without psychiatric comorbidities and female alcoholic patients with personality disorders (PD) had significantly lower ($p<0.001$) platelet 5-HT concentration than female healthy subjects.

When male alcoholic patients were subdivided according to the comorbid depression, they had significantly lower ($P=0.004$) platelet 5-HT concentration than male alcoholic patients without psychiatric comorbidities. Platelet 5-HT concentration in male alcoholic patients with comorbid depression was significantly ($p<0.001$) lower compared to values in healthy men, or to values in patients with depression. Depressive male patients had had significantly ($p=0.014$) lower platelet 5-HT concentration than healthy men (Figure 9).

Platelet 5-HT concentration differed significantly ($F=15.68$; $df=3, 182$; $p<0.001$, ANOVA) between 140 female healthy subjects, 16 female alcoholic patients without psychiatric comorbidities, 7 female alcoholics with depression, and 23 female patients with depression (Figure 10).

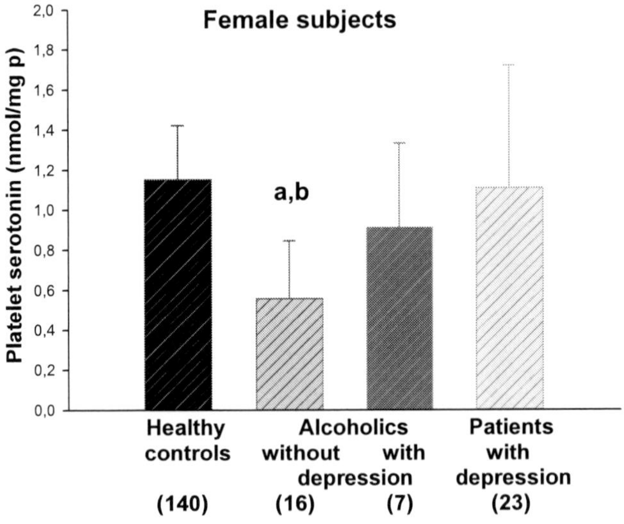

Figure 10. Platelet serotonin concentrations in female alcoholics with depression, female alcoholic patients without any psychiatric comorbidities, female patients with depression and healthy female controls. Each column represents mean ± SD. Number of subjects is given in brackets. [a] $p= 0.02$ vs. female alcoholics with depression; [b] $p<0.001$ vs. healthy controls (Student Newman Keuls test).

Female alcoholic patients with comorbid depression had significantly higher ($p=0.02$) platelet 5-HT concentration than female alcoholic patients without psychiatric comorbidities. Female alcoholic patients with comorbid depression had similar ($p=0.17$) values of platelet 5-HT as female patients with depression.

Female alcoholic patients without psychiatric comorbidities had significantly lower ($p<0.001$) platelet 5-HT concentration than female healthy subjects. Female patients with depression had similar ($p=0.56$) values of platelet 5-HT as female healthy subjects (Figure 10).

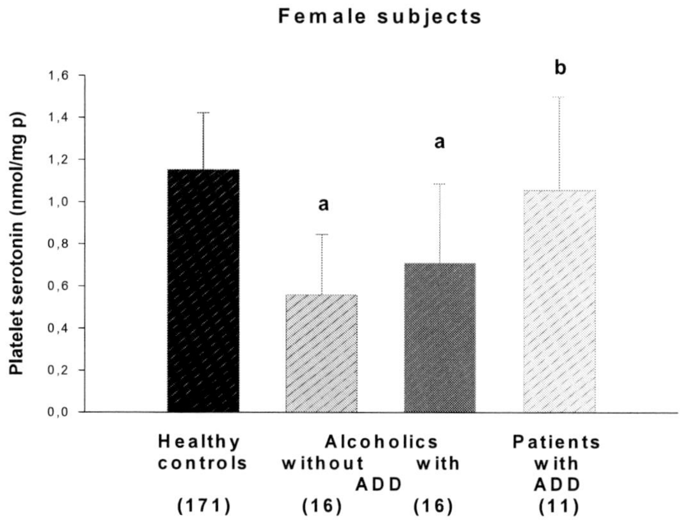

Figure 11. Platelet serotonin concentrations in female alcoholic patients with anxious-depressive disorder (ADD), female alcoholic patients without any psychiatric comorbidities, female patients with ADD and healthy female controls. Each column represents mean ± SD. Number of subjects is given in brackets. [a] $p<0.001$ vs. healthy women; [b]$p=0.002$ vs. female alcoholic patients with anxious-depressive disorder (ADD), (Student Newman Keuls test).

3.8. Platelet 5-HT Concentration in Female Alcoholic Patients with Anxious-Depressive Disorders, Alcoholic Patients without Psychiatric Comorbidities, Patients with Anxious-Depressive Disorders and Healthy Controls

Platelet 5-HT concentration differed significantly ($F=31.59$; $df=3, 210$; $p<0.001$, ANOVA) between 171 female healthy subjects, 16 female alcoholic patients without psychiatric comorbidities, 16 female alcoholics with anxious-depressive disorder (ADD), and 11 female patients with anxious-depressive disorder (ADD), (Figure 11).

Female alcoholic patients without ADD and female alcoholic patients with comorbid anxious-depressive disorder (ADD) had significantly lower ($p=0.001$) platelet 5-HT concentration than healthy control women. Female patients with anxious-depressive disorder had significantly lower ($p=0.002$) values of platelet 5-HT than healthy female controls (Figure 11).

3.9. Platelet MAO Activity in Alcoholic Patients and Healthy Controls

Platelet MAO activity was determined in 124 male alcoholic patients and in 42 female alcoholic patients and these values were compared with values in 225 male healthy subjects and in 78 female healthy control subjects.

Platelet MAO activity in 124 male and 42 female alcoholic patients and in 225 male and 124 female healthy controls differed significantly ($F=26.27$; $df=3, 465$; $p<0.001$, ANOVA).

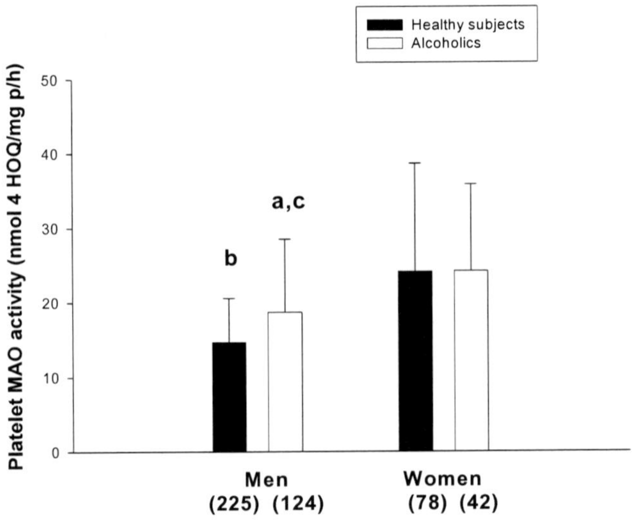

Figure 12. Platelet MAO activity in male and female healthy controls, and in alcoholic patients. Each column represents mean ± SD. Number of subjects is given in brackets. [a] $p=0.003$ vs. female alcoholics; [b] $p<0.001$ vs. healthy women; [c] $p<0.001$ vs. healthy men (Student Newman Keuls test).

Sex differences in platelet MAO activity were found, and platelet MAO activity was significantly ($p<0.001$) increased by 23% in female alcoholics, and by 39% in healthy women, when compared to corresponding male groups (Figure 12). Platelet MAO activity was not significantly ($p=0.973$) different between female healthy and alcoholic subjects. On the other hand, platelet MAO activity was significantly ($p<0.001$) higher in male alcoholics than in male healthy subjects.

3.10. Platelet MAO Activity in Alcoholic Patients and Healthy Controls Subdivided According to the Smoking Status into Smokers and Nonsmokers

Platelet MAO activity differed significantly ($F=3.11$; $df=3, 177$; $p<0.028$, ANOVA) between 139 male and 42 female alcoholic patients who were divided into smokers and nonsmokers (Figure 13), however these differences were due to the slightly higher platelet MAO activity in female alcoholics. Namely, nonsmokers had slightly higher platelet MAO activity than smokers, and this trend can be seen in male and female alcoholic patients, but this difference did not reach the level of statistical significance in multiple comparison test (Figure 13).

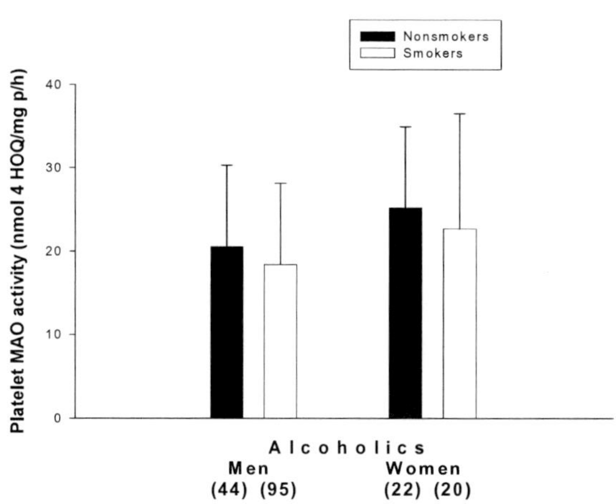

Figure 13. Platelet MAO activity in male and female alcoholic patients, who were divided into smokers or nonsmokers. Each column represents mean ± SD. Number of subjects is given in brackets.

Within healthy subjects, there were significant differences (F=33.39; df=3, 299; p<0.001, ANOVA) in platelet MAO activity in 225 male and 78 female healthy subjects, who were divided into smokers or nonsmokers (Figure 14).

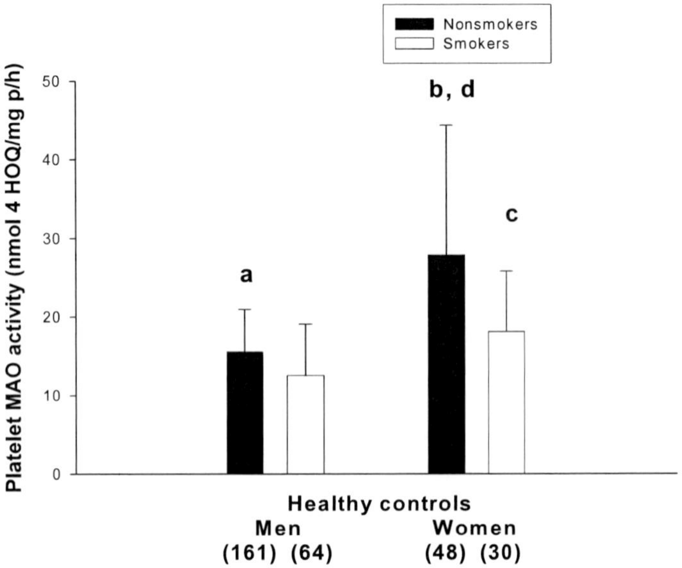

Figure 14. Platelet MAO activity in male and female healthy controls who were divided into smokers and nonsmokers. Each column represents mean ± SD. Number of subjects is given in brackets. [a] p=0.018 vs. healthy male smokers; [b] p<0.001 vs. healthy female smokers; [c]p<0.001 vs. healthy male smokers; [d]p<0.001 vs. healthy male nonsmokers (Student Newman Keuls test).

Platelet MAO activity was significantly (p<0.001) higher in female control subjects when compared to male control subjects, and this sex related difference was observed in smokers and nonsmokers. Nonsmoking healthy women had the highest platelet MAO activity, which was significantly increased when compared to values in healthy female smokers (p<0.001), or to healthy male nonsmokers (p<0.001). Healthy female smokers had significantly (p<0.001) higher platelet MAO activity than healthy male smokers. Healthy male nonsmokers had significantly (p<0.001) higher platelet MAO activity than healthy male smokers (Figure 14).

3.11. Platelet MAO Activity in Male Alcoholic Patients and Healthy Controls Subdivided According to the Number of Cigarettes Smoked Per Day

Platelet MAO activity was significantly different ($F=4.94$; $df=7, 226$; $p<0.001$, ANOVA) between male healthy controls and male alcoholic patients who were divided according to the number of cigarettes smoked per day (Figure 15).

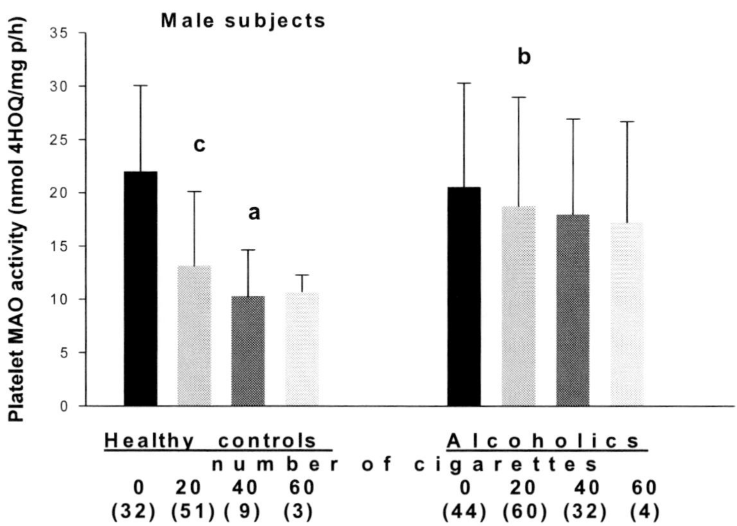

Figure 15. Platelet MAO activity and the number of cigarettes/day smoked in male alcoholics and healthy male controls. Each column represents mean ± SD. Number of subjects is given in brackets. [a] $p=0.010$ vs. healthy nonsmokers; [b] $p=0.005$ vs. healthy controls smoking 20 cigarettes/day; [c] $p<0.001$ vs. healthy nonsmokers (Student Newman Keuls test).

Platelet MAO activity showed a significant fall that was conversely related to number of cigarettes smoked per day. The highest MAO activity was found in male healthy nonsmokers, and platelet MAO activity declined significantly ($p<0.001$) in male healthy persons who smoked 20 cigarettes per day. This fall was significant ($p<0.01$) in healthy men who smoked 40 cigarettes per day. Platelet MAO activity of healthy men who smoked 60 cigarettes per day was also reduced, but due to only 3 subjects, this reduction was not significant.

Alcoholic male patients who did not smoke had similar (p=0.485) values of platelet MAO as the corresponding healthy male nonsmoking subjects. Male alcoholics who smoked 20 cigarettes per day had significantly (p<0.001) higher platelet MAO activity than healthy men who smoked 20 cigarettes per day (Figure 15).

When female subjects were divided according to the number of cigarettes smoked per day (Figure 16), their platelet MAO activity was significantly different (F=3.02; df=7, 115; p<0.021, ANOVA). Healthy women who smoked 20 cigarettes per day had significantly (p<0.001) lower platelet MAO activity than healthy nonsmoking women. A decreasing trend, conversely related to smoking, was found in alcoholic women, but this trend, i.e. low platelet MAO activity in alcoholic women who smoked 20, and lower platelet MAO activity in alcoholic women who smoked 40 cigarettes per day, compared to nonsmoking alcohol women, did not reach the level of statistical significance in multiple comparison test (Figure 16).

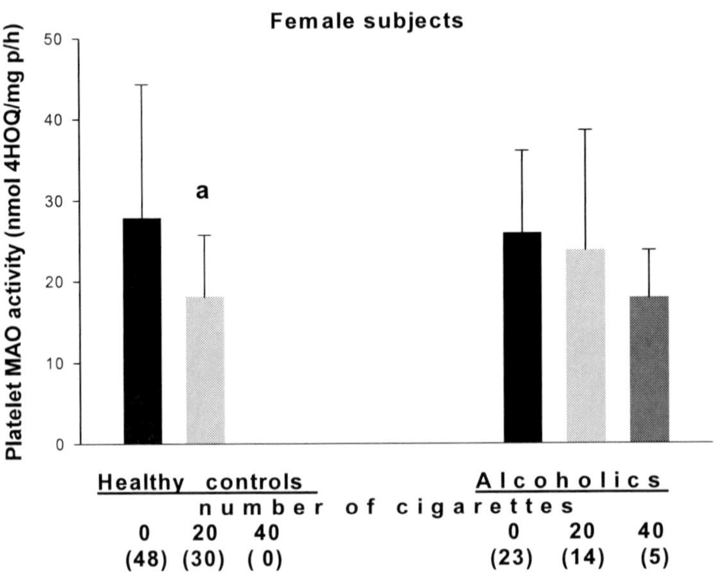

Figure 16. Platelet MAO activity and the number of cigarettes/day in female alcoholic patients and healthy female controls. Each column represents mean ± SD. Number of subjects is given in brackets. [a] p=0.009 vs. healthy nonsmokers (Student Newman Keuls test).

3.12. Platelet MAO Activity in Male and Female Alcoholic Patients and Healthy Controls Subdivided According to the Smoking Status and Different Psychiatric Comorbidities

Platelet MAO activity differed significantly ($F=6.78$; $df=5,341$; $p<0.001$, ANOVA) between male healthy controls and male alcoholic patients who were divided into patients without any psychiatric comorbidity and those with comorbid psychiatric diagnoses and smoking status (Figure 17).

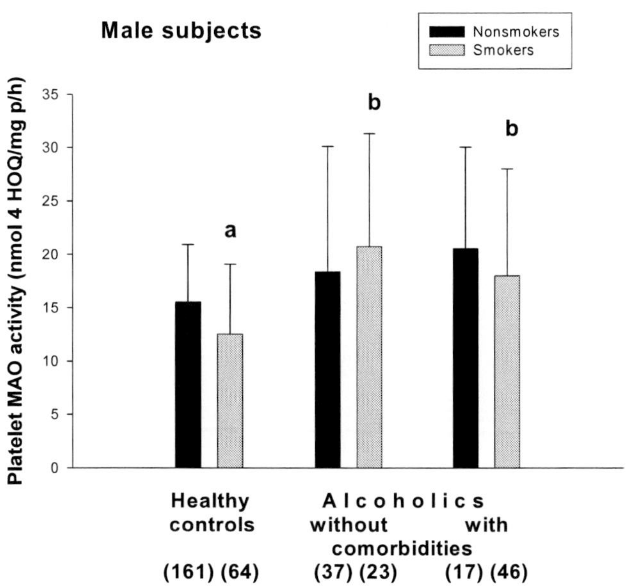

Figure 17. Platelet MAO activity in male alcoholic smokers and nonsmokers, without or with comorbidities, and in healthy male controls. Each column represents mean ± SD. Number of subjects is given in brackets. [a] $p=0.009$ vs. healthy male nonsmokers; [b] $p<0.001$ vs. healthy male smokers (Student Newman Keuls t-test).

Platelet MAO activity was significantly ($p=0.009$) lower in male healthy smokers than in healthy nonsmoking men. Within alcoholic male subjects, platelet MAO activity was significantly higher ($p<0.001$) in male alcoholics without psychiatric comorbidities or alcoholics with different psychiatric comorbidities than in healthy male smokers (Figure 17).

The comorbid psychiatric conditions did not affect significantly ($p=0.33$) platelet MAO activity within male nonsmoking alcoholic patients, or within male alcoholic ($p=0.47$) smokers (Figure 17).

Platelet MAO activity was not significantly different ($F=2.17$; $df=5,114$; $p=0.062$, ANOVA) between female healthy controls and female alcoholic patients who were divided into those without any psychiatric comorbidity and those with comorbid psychiatric diagnoses (Figure 18).

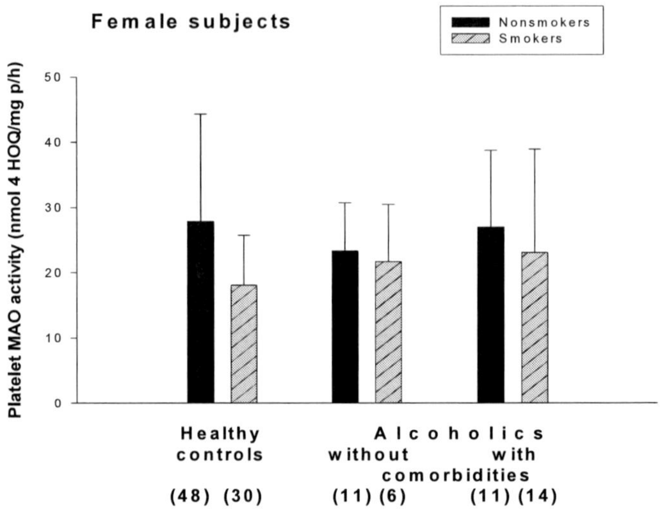

Figure 18. Platelet MAO activity in female alcoholic patients without or with comorbidities, and healthy female controls. Each column represents mean ± SD. Number of subjects is given in brackets.

3.13. Platelet MAO Activity in Male Alcoholic Patients and Healthy Controls Subdivided According to the Comorbid PTSD and Smoking Status

Platelet MAO activity differed significantly ($F=6.59$; $df=6,343$; $p<0.001$, ANOVA) between male healthy controls and male alcoholic patients who were divided into those with comorbid PTSD, alcoholic patients without any psychiatric comorbidity, and male PTSD patients (Figure 19). However, comorbid PTSD diagnosis within alcoholic patients did not affect significantly platelet MAO activity.

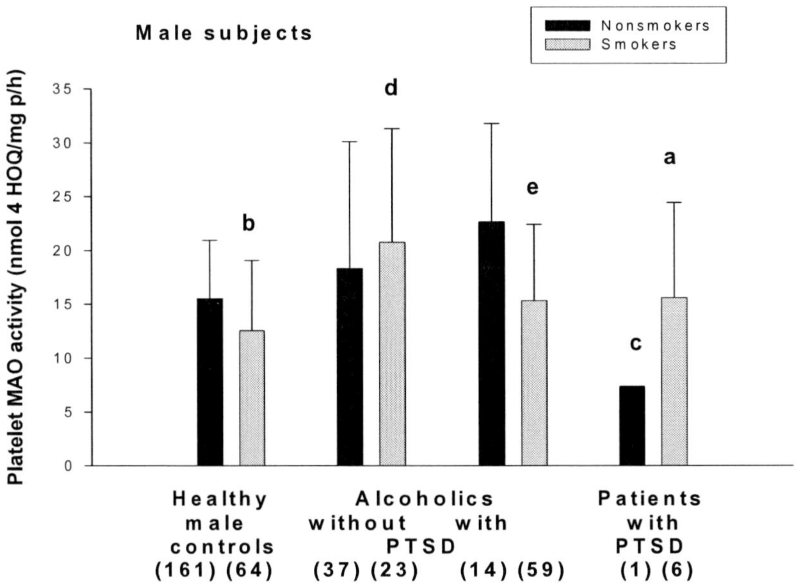

Figure 19. Platelet MAO activity in male alcoholics without or with posttraumatic stress disorder (PTSD), patients with PTSD and healthy male controls. Each column represents mean ± SD. Number of subjects is given in brackets: [a] p=0.018 vs. healthy smokers; [b] p=0.016 vs. healthy nonsmokers; [c]p=0.007 vs. healthy nonsmokers; [d]p<0.001 vs. healthy smokers; [e]p<0.001 vs. alcoholic patients with PTSD who were nonsmokers (Student Newman Keuls test).

Smoking status significantly lowered platelet MAO activity in healthy male smokers (p=0.016) and in alcoholic male patients with PTSD who smoked (p=0.001). Male patients with PTSD who did not smoke had significantly (p=0.007) higher platelet MAO activity than healthy male nonsmokers. Within alcoholic male subjects, platelet MAO activity was significantly higher (p<0.001) in male alcoholics without psychiatric comorbidities than in healthy male smokers (Figure 19).

3.14. Platelet MAO Activity in Male Alcoholic Patients and Healthy Controls Subdivided According to the Comorbid Schizophrenia and Smoking Status

Platelet MAO activity differed significantly ($F=6.40$; $df=5,38$; $p<0.001$, ANOVA) between male healthy controls and male alcoholic patients who were divided into those with comorbid schizophrenia, alcoholic patients without any psychiatric comorbidity, and male schizophrenic patients (Figure 20). However, comorbid schizophrenia within alcoholic patients did not affect significantly platelet MAO activity.

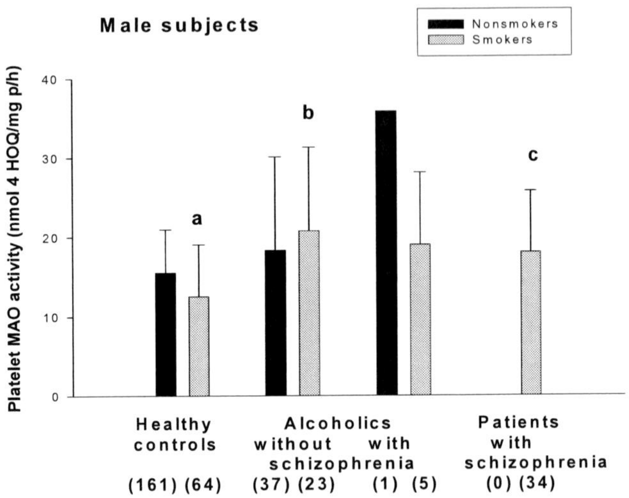

Figure 20. Platelet MAO activity in male alcoholics without or with schizophrenia, patients with schizophrenia, and healthy male controls. Each column represents mean ± SD. Number of subjects is given in brackets. [a] $p=0.006$ vs. healthy nonsmokers; [b] $p<0.001$ vs. healthy smokers; [c] $p<0.001$ vs. healthy smokers (Student Newman Keuls test).

Smoking significantly decreased platelet MAO activity in healthy controls ($p=0.006$). Schizophrenic male patients who smoked had significantly higher ($p<0.001$) platelet MAO activity than healthy male smokers. Male alcoholic patients with schizophrenia who smoked had similar ($p=0.96$) platelet MAO activity as schizophrenic patients who smoked. Platelet MAO activity was significantly higher ($p<0.001$) in male alcoholics without psychiatric comorbidities than in healthy male smokers (Figure 20).

3.15. Platelet MAO Activity in Male and Female Alcoholic Patients and Healthy Controls Subdivided According to the Smoking Status and Comorbid Personality Disorders

Platelet MAO activity differed significantly (F=5.03; df=6,321; p<0.001, ANOVA) between male healthy controls and male alcoholic patients who were divided into those with personality disorders, alcoholic patients without any psychiatric comorbidity, and male patients with personality disorders (Figure 21). However, comorbid personality disorders within alcoholic patients did not affect significantly platelet MAO activity.

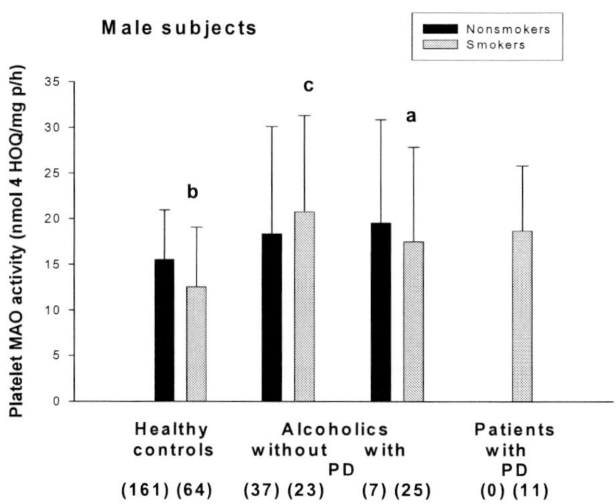

Figure 21. Platelet MAO activity in male alcoholics without or with personality disorder (PD), patients with PD, and healthy male controls. Each column represents mean ± SD. Number of subjects is given in brackets. [a] p=0.016 vs. healthy smokers; [b] p=0.008 vs. healthy nonsmokers; [c] p<0.001 vs. healthy smokers (Student Newman Keuls test).

Healthy smokers had significantly lower (p=0.008) platelet MAO activity than healthy nonsmokers. Male alcoholic patients with personality disorders who smoked had significantly higher (p=0.016) platelet MAO activity than healthy male smokers. Male alcoholic patients without psychiatric comorbidities who smoked had significantly higher (p<0.001) platelet MAO activity than healthy male smokers (Figure 21).

Platelet MAO activity differed significantly (F=4.77; df=6,105; p<0.001, ANOVA) between healthy women and female alcoholic patients who were divided into those with personality disorders, alcoholic women without any psychiatric comorbidity, and female patients with personality disorders (Figure 22). Within alcoholic patients, comorbid personality disorder did not affect significantly platelet MAO activity.

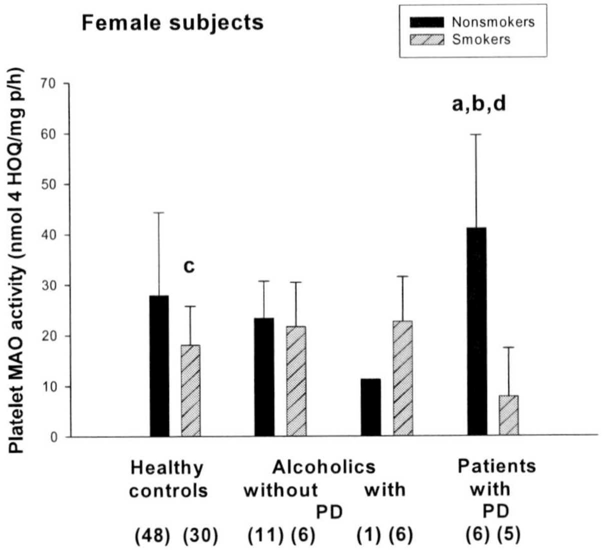

Figure 22. Platelet MAO activity in female alcoholics without or with personality disorder (PD), patients with PD, and healthy female controls. Each column represents mean ± SD. Number of subjects is given in brackets. [a] p=0.021 vs. healthy control nonsmokers; [b] p=0.023 vs. healthy controls smokers; [c] p=0.015 vs. healthy control nonsmokers; [d] p=0.001 vs. female patients with PD who smoked (Student Newman Keuls test).

Healthy female smokers had significantly lower (p=0.015) platelet MAO activity than healthy women who did not smoke. Female patients with personality disorders who were not smokers had significantly higher platelet MAO activity than healthy female smokers (p=0.021), than healthy female nonsmokers (p=0.032) and than female patients with personality disorders who smoked (p=0.001) (Figure 22).

3.16. Platelet MAO Activity in Male and Female Alcoholic Patients and Healthy Controls Subdivided According to the Comorbid Depression and Smoking Status

Platelet MAO activity differed significantly ($F=7.25$; $df=6,315$; $p<0.001$, ANOVA) between healthy men and male alcoholic patients who were divided into those with comorbid depression, alcoholic men without any psychiatric comorbidity, and male depressive patients (Figure 23).

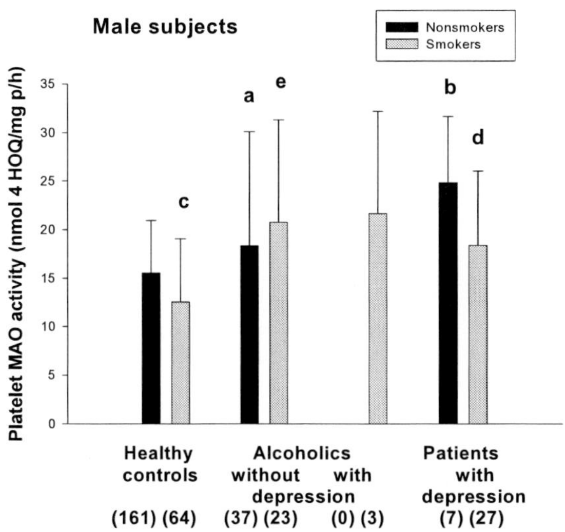

Figure 23. Platelet MAO activity in male alcoholics without or with depression, patients with depression, and healthy male controls. Each column represents mean ± SD. Number of subjects is given in brackets. [a] $p=0.035$ vs. healthy nonsmokers; [b] $p=0.013$ vs. healthy nonsmokers; [c] $p=0.006$ vs. healthy nonsmokers; [d] $p=0.003$ vs. healthy smokers; [e] $p<0.001$ vs. healthy smokers (Student Newman Keuls test).

Healthy male nonsmokers had significantly higher ($p=0.006$) platelet MAO activity than healthy male smokers. Comorbid depression did not affect significantly platelet MAO between alcoholic patients. Male patients without comorbid psychiatric diagnoses who were not smokers had significantly higher platelet MAO activity than healthy male nonsmokers ($p=0.001$), and male patients without comorbid psychiatric diagnoses who smoked had significantly higher platelet MAO activity than healthy male smokers ($p=0.035$). Male depressive patients who did not smoke had significantly higher platelet MAO activity than healthy male nonsmokers (p=0.013), while depressive male smoking patients had significantly higher platelet MAO activity than healthy male smokers (p=0.003) (Figure 23).

Platelet MAO activity was similar and did not differ significantly (F=1.84; df=6,123; p=0.097, ANOVA) between healthy women and female alcoholic patients who were divided into those with comorbid depression, alcoholic female patients without any psychiatric comorbidity, and female depressive patients (Figure 24).

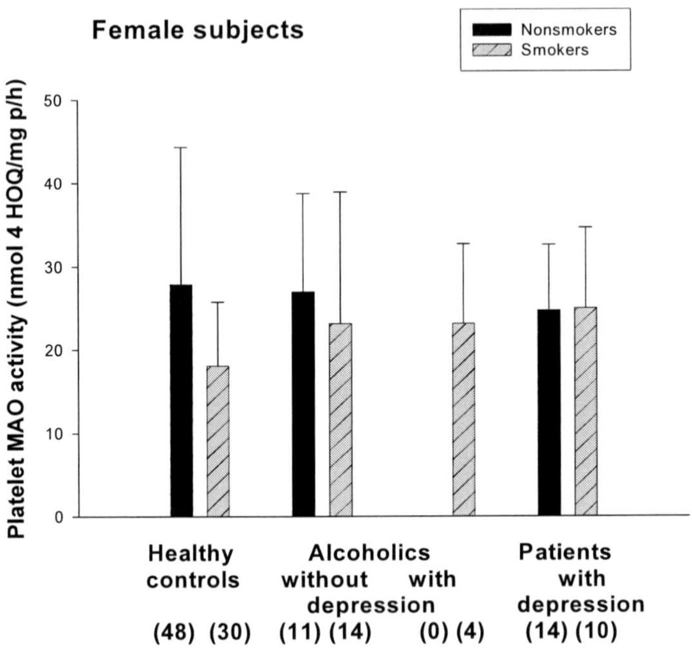

Figure 24. Platelet MAO activity in female alcoholics without or with depression, patients with depression, and healthy female controls. Each column represents mean ± SD. Number of subjects is given in brackets.

3.17. Platelet MAO Activity in Female Healthy Controls and in Female Alcoholic Patients Subdivided According to the Comorbid Anxious-Depressive Disorder and Smoking Status

Platelet MAO activity was alike and did not differ significantly (F=1.87; df=7,114; p=0.082, ANOVA) between healthy women and female alcoholic patients who were divided into those with comorbid anxious-depressive disorder, alcoholic female patients without any psychiatric comorbidity, and female patients with anxious-depressive disorder (Figure 25).

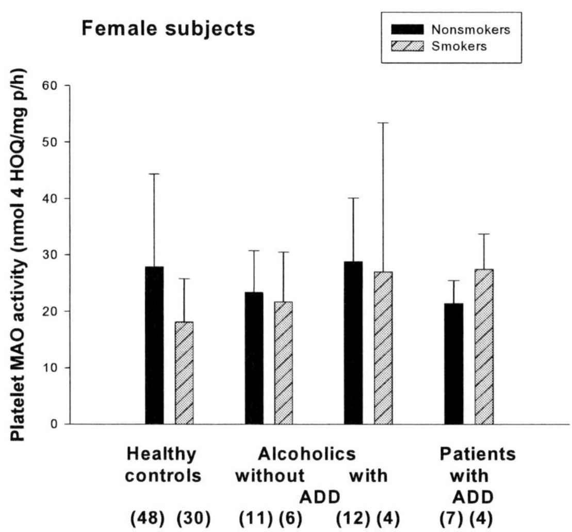

Figure 25. Platelet MAO activity in female alcoholics without or with anxious-depressive disorder (ADD), patients with ADD, and healthy female controls. Each column represents mean ± SD. Number of subjects is given in brackets.

Age of the patients did not differ significantly (t= 1.491, df=191, Student's t-test), Table 3. When age was compared between alcoholic patients and healthy control subjects, alcoholic subjects and healthy control subjects differed significantly (F=22.04; df=3, 322; p<0.001) in age. Both male and female alcoholics were significantly (p<0.001) older than male or female healthy subjects. No significant (p>0.05) difference in age was found between male or female alcoholic subjects, as well as between male or female healthy subjects.

There was no significant correlation between platelet MAO activity and age in female alcoholics ($r = 0.033$, $p = 0.836$), male alcoholics ($r = 0.065$; $p = 0.439$), healthy men ($r = 0.002$, $p = 0.990$) or healthy women ($r = 0.188$; $p = 0.252$).

Table 3. Demographic data for male and female alcoholic patients. Results are expressed as mean ± SD

	Alcoholics	
	Men (152)	Women (41)
Age (years)	49.1 ± 10,1	51.8 ±10.4
Age at first drinking (years)	17.6 ± 4.3	20.0 ± 6.5
Age at strong drinking (years)	29.1 ± 8.8	34.6 ± 10.5
Duration of drinking (years)	11.1 ± 7.9	8.2 ± 7.7
Age at first hospitalization (years)	40.6 ± 11.3	45.1 ± 12.5
Number of hospitalizations	6.9 ± 9.8	6.3 ± 9.1
Number of suicide attempts	1.8 ± 0.4	1.8 ± 0.5
HAMA (scores)	25.5 ± 7.8	27.6 ± 8.7
HAMD (scores)	21.5 ± 7.0	23.8 ± 7.4
CAGE (scores)	3.8 ± 0.5	3.7 ± 0.6

Table 4. Two-way ANOVA for the platelet MAO activity (main effect) in subjects with different diagnosis (healthy controls, alcoholics), sex (female, male) and smoking status (smokers, nonsmokers)

Source of variance	df	F-value	Significance (p)
MAIN EFFECT	1,2	21.755	0.000
Sex	1,2	34.076	0.000
Smoking	1,2	8.545	0.004
2-FACTOR INTERACTIONS:			
Sex x smoking	1,2	4.854	0.028
MAIN EFFECT	1.2	4.962	0.008
Diagnosis	1.2	1.365	0.243 n.s.
Smoking	1,2	9.853	0.002
2-FACTOR INTERACTIONS:			
Diagnosis x smoking	1,2	0.544	0.469 n.s.
MAIN EFFECT	1.2	17,449	0.000
Diagnosis	1.2	0.921	0.348 n.s.
Sex	1.2	34.821	0.000
2-FACTOR INTERACTIONS:			
Diagnosis x sex	1.2	1.922	0.166 n.s.

n.s. = non-significant; MAO= monoamine oxidase.

Two-way ANOVA revealed the significant effect of sex (male/female) and smoking (smokers/nonsmokers), but no significant effect of diagnosis (alcoholism /no alcoholism) on platelet MAO activity in healthy and alcoholic men and women (Table 4). The interaction between sex and smoking was significant, but the interactions between sex and diagnosis, or diagnosis and smoking were not significant (Table 4).

Within alcoholic male and female subjects, the effect of sex, smoking and different comorbid diagnoses (personality disorder, PTSD, anxious-depressive disorder, major depression, schizophrenia) were evaluated on platelet MAO activity (two-way ANOVA). It showed a significant effect of sex, and no significant effect of smoking or different comorbid diagnoses, on platelet MAO activity. No significant interaction between sex and smoking, or different comorbidities and smoking, was detected. The interaction between sex and different comorbidities could not be done due to the different number of comorbidities between male or female alcoholics (Table 5).

Table 5. Two-way ANOVA for the platelet MAO activity (main effect) in male and female (sex) alcoholic subjects without and with different comorbid diagnoses (personality disorder, anxious-depressive disorder, depression, PTSD, schizophrenia) and smoking status (smokers, nonsmokers)

Source of variance	df	F-value	Significance (p)
MAIN EFFECT	1,2	5.069	0.007
Sex	1,2	6.912	0.009
Smoking	1,2	1.408	0.237 n.s.
2-FACTOR INTERACTIONS:			
Sex x smoking	1,2	0.054	0.818 n.s.
MAIN EFFECT	1.2	1.544	0.166 n.s.
Comorbid diagnoses	1.2	1.219	0.302 n.s.
Smoking	1,2	1.846	0.176 n.s.
2-FACTOR INTERACTIONS:			
Comorbid diagnoses x smoking	1,2	1.099	0.358 n.s.

n.s. = non-significant; MAO=monoamine oxidase; PTSD=posttraumatic stress disorder .

Chapter 4

Discussion

4.1. Platelet 5-HT Concentration in Alcoholic Patients and Healthy Controls

The results from the present study, obtained on a large groups of male and female drug-free alcoholic patients (i.e. patients with withdrawal syndrome related to alcohol dependency) and healthy subjects, show that alcoholic patients had significantly lower platelet 5-HT concentration than the corresponding healthy controls, and confirm the data from our previous (Pivac et al., 2004), and other (Bailly et al., 1990; Bailly et al., 1993; Schmidt et al., 1997; Borcsiczky et al., 1996) studies. One study did not detect any significant differences in platelet 5-HT content between alcoholic and control subjects (Szalay et al., 2001), however, this study included patients with alcohol-induced liver cirrhosis.

4.2. Sex Differences in Platelet 5-HT Concentration in Alcoholic Patients and Healthy Controls

Our data confirmed previous finding, showing the reduced platelet 5-HT in alcoholism, and extend them with the inclusion of large numbers of both male and female patients. With the exception of our previous study (Pivac et al., 2004), the other studies included only male patients (Patkar et al., 2003), or combined male and female patients (Schmidt et al., 1997), or did not detect any male/female differences in platelet 5-HT concentration (Bailly et al., 1990; Bailly et al., 1993). In the present study both male and female alcoholic subjects had significantly lower platelet 5-HT concentration, than male or female healthy control subjects. In line with the sex-related differences in some peripheral serotonergic measures (5-HT transporter gene, whole blood 5-HT level, and platelet 5-HT binding) in families with alcoholism (Stoltenberg et al., 2002), and in platelet 5-HT concentration, found in healthy subjects (Oxenkrug, 1979; Muck-Seler et al., 1996a; Muck-Seler et al., 1999; Pivac et al., 2001), or alcoholic patients (Pivac et al., 2004), we have found significant sex-related differences in platelet 5-HT concentration, both between male and female alcoholic patients, as well as between male and female healthy control subjects. Female alcoholic patients had significantly lower platelet 5-HT concentration than alcoholic male patients. The differences between studies might be explained by the differences in the number of patients (i.e. Bailly et al. (1990) had a small number (N=9) of female alcoholics). On the other hand, the same group did not find any significant sex-related differences in platelet 5-HT concentration between male and female alcoholic patients in the bigger groups of alcoholic patients (Bailly et al., 1993). Sex differences, found in healthy subjects (present study, Oxenkrug, 1979; Muck-Seler et al., 1996a; Muck-Seler et al., 1999; Pivac et al., 2001; Pivac et al., 2004), depressed (Oxenkrug, 1979; Muck-Seler et al., 1996a), suicidal (Muck-Seler et al., 1996a; Mueller-Oerlinghausen et al., 2004), or schizophrenic (Muck-Seler et al., 1999) patients, were also detected in alcoholic subjects. These sex-related differences are in line with the lower plasma free tryptophan, and lower 5-HT synthesis in the brain of female when compared to male healthy subjects (Nishizawa et al., 1997). The decreased rate of the 5-HT synthesis in healthy women might indicate that under stressful situations central 5-HT would decline more in women than in men, leading to vulnerability to different psychiatric disorders (Nishizawa et al., 1997). Reduced platelet 5-HT concentration in female subjects, found across different diagnoses and in healthy controls, does not appear to be related to the sampling of females in the different phases of the estrus cycle, or to estrogens levels in female subjects, since estro-progestative treatment did not change platelet and blood 5-HT concentration in female depressed patients (Guicheney et al., 1988), and estrogen replacement therapy did not affect significantly plasma 5-HT levels in postmenopausal women (Blum et al., 1996). In addition, we did not detect any significant changes in platelet 5-HT concentration between pre-menopausal and post-menopausal healthy, depressed or schizophrenic women (Muck-Seler et al., 2004). On the other hand, the opposite data came from the study which showed variations in plasma 5-HT concentrations during menstrual cycle, however, the data were obtained on a small number (N=19) of women in different menstrual cycles (Hindberg and Maesh, 1992) and used plasma and not platelet 5-HT levels. Therefore, the possible effect of estrus cycle on platelet 5-HT concentration might be excluded since there were no significant differences in the whole blood 5-HT levels in healthy women during follicular, ovulatory or luteal phase of the menstrual cycle (Rasgon et al., 2000), and the studies evaluating the possible effect of the menstrual cycle on platelet 5-HT concentration are not consistent (Leibenluft et al., 1994).

4.3. The Lack of Effect of Smoking on Platelet 5-HT Concentration in Alcoholic Patients and Healthy Controls

Platelet 5-HT concentration was not significantly different between male or female alcoholic patients and healthy subjects who were divided according to the smoking status into smokers and nonsmokers in the present study and in our previous (Pivac et al., 2004) study. These data suggest that smoking status, per se, does not significantly alter platelet 5-HT concentration in male or female alcoholic or healthy subjects. This finding is in contrast with the increasing effect of smoking on 5-HT concentration in alcoholic patients (Schmidt et al., 1997). However, the same authors did not detect the significant effect of smoking on platelet 5-HT concentration in healthy control subjects (Schmidt et al., 1997). The lack of any significant effect of smoking on platelet 5-HT concentration in alcoholic male or female subjects, or in healthy subjects, and the fact that alcoholic patients, who smoked 20-60 cigarettes per day, had similar platelet 5-HT concentration as the corresponding non-smoking alcoholic patients, which were reduced compared to healthy control subjects, does not support the hypothesis that smoking "normalizes" decreased platelet 5-HT concentrations in alcoholic patients (Schmidt et al., 1997), and excludes a smoking as a confounding factor.

4.4. The Effect of Various Psychiatric Comorbidities on Platelet 5-HT oncentration in Alcoholic Patients and Healthy Controls

In line with the proposition that various psychiatric comorbidities associated to alcohol abuse might modify platelet 5-HT concentration (Bailly et al., 1993; Schmidt et al., 1997), male and female alcoholic patients with different psychiatric comorbidities (personality disorder, PTSD, anxious-depressive disorder, major depression, schizophrenia) had different platelet 5-HT concentration. However, detailed statistical evaluation of these data showed that these significant differences in platelet 5-HT concentrations were due to the significantly higher platelet 5-HT concentration in alcoholic male subjects with comorbid PTSD, when compared to male alcoholic patients without PTSD. Alcoholic patients without any psychiatric diagnoses (i.e. "pure" alcoholic patients) had decreased platelet 5-HT concentration, and the comorbid PTSD significantly increased these values when compared to alcoholic patients without comorbid PTSD. However, these platelet 5-HT values were not significantly different from platelet 5-HT concentration in male patients with PTSD, or from healthy male subjects. Hence, our data suggest that the presence of comorbid PTSD only "normalized" already reduced platelet 5-HT concentration in alcoholic patients. Therefore, these alcoholic patients with PTSD had similar platelet 5-HT concentration as patients with PTSD, who were not alcoholics, and these results agree with the alike values of platelet 5-HT detected in war veterans with PTSD who were not alcoholics and the corresponding control subjects (Pivac et al., 2002; Muck Seler et al., 2003; Pivac et al., 2006a; Pivac et al., 2006b). In addition, the increased platelet 5-HT concentration in alcoholic male subjects with comorbid PTSD might be due to the undiagnosed presence of psychotic symptoms, which would increase platelet 5-HT concentration. Namely, we have repeatedly shown that the presence of psychotic symptoms in depression (Muck-Seler et al., 1996a; Pivac et al., 1997a), mania (Sagud et al., 2006), psychotic features in PTSD (Pivac et al., 2006a), or positive symptoms in schizophrenia (Jakovljevic et al., 1997; Pivac et al., 1997b) increase platelet 5-HT concentration.

In agreement with our previous data obtained on male (Muck-Seler et al., 1996a; Pivac et al., 1997a) and female nonpsychotic depressed patients (Muck-Seler et al., 1996a; Pivac et al., 2003; Muck-Seler et al., 2004; Muck-Seler et al., 2005), and in male alcoholic patients with depression (Pivac et al., 2004), who had decreased baseline platelet 5-HT concentration compared to corresponding healthy subjects, the presence of comorbid depression in male alcoholic patients significantly decreased platelet 5-HT concentration. Depressive alcoholic male patients had significantly lower platelet 5-HT values than male alcoholic patients without depression, or than non-alcoholic patients with depression, or than healthy control men. These data disagree with the similar platelet 5-HT levels detected between alcoholics with anxiety/depression, other comorbidities or no comorbidity (Schmidt et al., 1997). In addition, male psychiatric patients with depression had significantly lower platelet 5-HT concentration than healthy men, a result in line with our previous (Muck-Seler et al., 1996a; Pivac et al., 1997a) and other (Oxenkrug, 1979) data. These reductions in platelet 5-HT content might be associated with low concentrations of total plasma tryptophan concentration (Westenberg et al., 1982), and lower 5-HT synthesis (Rosa-Neto et al., 2004) in depressed patients than in healthy subjects, since a positive correlation between plasma tryptophan concentration and 5-HT synthesis rate was detected (Nishizawa et al., 1997). The discrepancies between the data showing either decreased (present and our previous studies), and unaltered (Franke et al., 2000) platelet 5-HT concentration in depressive patients are presumably due to the fact that depression, although a single psychiatric entity, differs when characterized with predominantly psychotic or predominantly with nonpsychotic symptoms (Jakovljevic et al., 1997; Leyton et al., 1995). Therefore, biological markers are also different in the two subtypes of depression (i.e. nonpsychotic and psychotic depression). Hence, our data might suggest that alcoholic depressed patients had depression with predominantly nonpsychotic symptoms, and consequently lower platelet 5-HT concentration than other alcoholic subjects without depression, or than healthy control subjects.

In contrast to male alcoholic patients, female alcoholic patients with comorbid depression had significantly higher platelet 5-HT concentration than female alcoholic patients without depression. However, it should be stressed out that their platelet 5-HT concentration did not differ from the values in depressed female psychiatric patients or healthy control women. This result is in line with the previous data when platelet 5-HT concentration was determined in depressed patients who were not divided into psychotic or nonpsychotic subtype (Jakovljevic et al., 1997; Muck-Šeler et al., 2002). In the present study the opposite finding of the higher platelet 5-HT concentration, compared to male alcoholic subjects, might suggests that female alcoholic patients and female depressed patients had depression that was characterized with predominantly psychotic symptoms, hence the presence of these psychotic symptoms partly "normalized" decreased platelet 5-HT concentration in alcoholic female patients with comorbid depression (Muck-Seler et al., 1996a; Pivac et al., 1997a).

It has been proposed that different comorbidities, especially antisocial and impulse control disorders, might affect platelet 5-HT content, and in line with that, antisocial alcoholics (Schmidt et al., 1997), and alcoholics with impulse control disorders (Bailly et al., 1993) had increased platelet 5-HT content. On the other hand, the same group (Schmidt et al., 1997) found no significant differences in platelet 5-HT levels between alcoholic patients who had anxiety/depression, alcoholics with other comorbidities or without comorbidity (Schmidt et al., 1997). Hence, it appears that there is no consensus related to the possible effect of different comorbidities on platelet 5-HT concentration in alcoholic patients (Schmidt et al., 1997). In contrast to the above cited data (Bailly et al., 1993; Schmidt et al., 1997), our male and female alcoholic patients with personality disorders or anxious-depressive disorder, and our male alcoholic patients with schizophrenia, had similar platelet 5-HT concentration as their comparative corresponding alcoholic patients without these psychiatric comorbidities. It seams that acute intoxication with alcohol, and the withdrawal syndrome related to alcohol dependency, observed in our alcoholic patients, blunted the possible influence of other psychiatric comorbidities on platelet 5-HT concentration.

Platelet 5-HT concentration was not decreased in alcoholic patients because of the older age. Healthy older women were reported to have increased platelet 5-HT and 5-HIAA (Kumar et al., 1998), and in our study alcoholic women were slightly but not significantly older than alcoholic men, but their platelet 5-HT was lower, and was not significantly correlated to age. Therefore, age did not affect significantly platelet 5-HT concentration, since age was not significantly associated with platelet 5-HT concentration in healthy subjects (present study, Muck-Seler et al., 1996b; Muck-Seler et al., 2003; Pivac et al., 2001; Pivac et al., 2003; Pivac et al., 2004), or in alcoholics (present study, Bailly et al., 1993; Pivac et al., 2004) or with platelet 5-HT uptake in alcoholic subjects (Javors et al., 2000).

Besides sex-related differences, the differences in platelet 5-HT concentration might be attributed to the heterogenity of the alcoholic patients, i.e. the inclusion of different diagnostic criteria, i.e. different population of alcoholics, such as abstinent alcoholics, or patients with acute alcohol intoxication, patients in the withdrawal syndrome, patients with alcoholism type I or II (Cloninger, 1987). All these variables might affect 5-HT measures. To avoid these confounding factors, in our study platelet 5-HT concentration was determined 12 h after the admission of alcoholic patients into the hospital, i.e. in sober alcoholic dependent patients, who were admitted due to the acute alcohol intoxication. Our data suggest that platelet 5-HT concentration is decreased in different groups of alcoholic patients regardless of the time-period after alcohol intoxication, or regardless of the acute alcohol consumption, since platelet 5-HT levels were decreased in acutely-intoxicated alcoholics (Schmidt et al., 1997), in alcoholics during withdrawal (present study, Bailly et al., 1990, 1993) or in alcoholics after 2 weeks of abstinence (Bailly et al., 1993). These data are also in line with the data showing a reduced plasma 5-HT content in alcohol dependent subjects hospitalized for detoxification from alcohol, who had decreased plasma 5-HT levels on admission, and after 1, 7 and 14 days during withdrawal (Patkar et al., 2003). The reduced peripheral indices of 5-HT function in withdrawal might be due to the fact that alcohol consumption stimulates 5-HT turnover, therefore the intake of alcohol might be related to the attempt to compensate for the deficient 5-HT function (see Patkar et al., 2003). In addition, the withdrawal is related to reduction in the 5-HT firing rates and reduced 5-HT release (Bailey et al., 2000), and therefore reduced platelet 5-HT content does not «normalize» during different periods of withdrawal (Bailly et al., 1990, 1993, 2000; Arranz et al., 1999). Reduced platelet 5-HT concentration in withdrawal might be related to the deficits in the 5-HT function in the brain, or the lowered plasma tryptophan availability, which was detected in early-onset alcoholism (Buydens-Branchey et al., 1989), or the decreased plasma tryptophan levels (Branchey et al., 1981). The decreased function of central 5-HT has been found in the post-mortem brain tissue from alcoholics, with decreased 5-HT transporter binding in the hippocampus (Gross-Isseroff and Biegon, 1988; Chen et al., 1991), and reduced binding in the 5-HT1A receptors of alcoholics compared to control subjects (Dillon et al., 1991). The other possible explanation is that these alcoholic patients had severely impaired 5-HT synthesis in the entherochromaffine cells of the gastrointestinal tract (LeMarquand et al., 1994), and therefore reduced platelet 5-HT concentrations.

To prevent the possible influence of the diurnal rhythm on the platelet 5-HT concentration, subjects were sampled always between 8.00-9.00 a.m. (Wirtz-Justice et al., 1977), and to prevent the possible seasonal influence on peripheral 5-HT measures (Muller-Oerlinghausen et al., 2004), we sampled alcoholic patients and healthy controls simultaneously. In addition, we have previously shown that male healthy, depressed and schizophrenic subjects (Jakovljevic et al., 1997) had similar platelet 5-HT concentration if sampled during different seasons.

Our findings of the decreased platelet 5-HT concentrations, obtained in the large groups of male and female ethnically homogenous drug-free alcoholic patients, in the withdrawal syndrome related to alcohol dependency, might support the hypothesis that platelet 5-HT concentration, among other markers, might be used as a limited peripheral model for the central serotonergic synaptosomes (Da Prada et al., 1988; Meszaros et al., 1998) in alcoholism. Our data are in line with the altered indicators of the central 5-HT activity (low activity of the central 5-HT neurotransmission (LeMarquand et al 1994), reduced 5-HT transporter binding (Gross-Isseroff and Biegon, 1988; Chen et al., 1991), reduced binding in the 5-HT1A receptors (Dillon et al., 1991), reduced levels of CSF 5-HIAA levels (Ballenger et al., 1979; Banki, 1981) which all indicate a central 5-HT disturbance in alcoholism. It has been discussed recently that peripheral 5-HT markers are associated to a single psychological dysfunction (Mueller-Oerlinghausen et al., 2004; van Praag, 2004), and in line with that hypothesis, our data suggest that platelet 5-HT concentration is related to specific withdrawal state in alcoholic patients.

4.5. Platelet MAO Activity in Alcoholic Patients and in Healthy Subjects

The present study showed that platelet MAO activity, determined in the large groups of drug-free male and female subjects with alcohol dependency, differed significantly when compared to sex-matched healthy controls. Platelet MAO activity was similar between female alcoholic patients and healthy women, but it was significantly higher in alcoholic male patients than in healthy men. This was an unexpected finding, since most studies showed either lower (Oreland et al., 2002; Oreland, 2004) or unchanged (Anthenelli et al., 1998; Farren et al., 1998; Verkes et al., 1998; Whitfield at al., 2000; Pivac et al., 2005) platelet MAO activity in alcoholic subjects, when compared to healthy subjects. Since our alcoholic subjects have been screened by the two experienced psychiatrists before admission into Psychiatric Hospital Vrapče, Center for Alcoholism and Other Addictions, and evaluated by the use of structured clinical interview, different disorders such as pernicious anemia, Alzheimer's disease, Parkinson's disease and Huntington’s chorea, that might increase platelet MAO activity (Oreland, 2004), were excluded.

In our study platelet MAO activity was not altered between female drug-free alcoholic subjects during withdrawal, when compared to female control subjects, confirming the hypothesis that peripheral serotonergic markers are linked to a single psychological dysfunction (Muller-Oerlinghausen et al., 2004), or to a temperament (Mezzich et al., 1994), and not to the psychiatric diagnosis (such as alcoholism). This finding agrees with the data showing similar platelet MAO activity in various types of alcoholism and in healthy controls (Anthenelli et al,. 1998; Farren et al., 1998; Verkes et al., 1998; Whitfield at al., 2000), or between alcoholic subjects who were admitted into hospital for a special alcohol detoxification, and those who were readmitted during 4 years period (Ponzer et al., 2002).

On the other hand, male alcoholic patients had significantly higher platelet MAO activity than their corresponding male healthy subjects. This result is in contrast with our data (Pivac et al., 2005) and other previous findings (Oreland et al., 2002; Oreland, 2004). The discrepancy might be explained by the fact that we included much bigger group of control male subjects in the present study, and detected a slight but significant difference in platelet MAO activity between male alcoholic and male healthy subjects. In support to this explanation, in our previous study (Pivac et al., 2005) we have noted a trend towards the higher platelet MAO activity in male alcoholic patients, when compared to male healthy controls, but due to the great dispersion of the data in healthy control subjects, this difference was not significant. However, our data do not agree with the reduced platelet MAO activity in alcoholic patients (Schmidt et al., 1997; Snell et al,. 2002; Oreland, 2004), or in intoxicated drunken drivers (Eensoo et al., 2003). These differences are presumably due to the methodological differences, diagnostic variability, and confounding effects of acute vs. chronic alcohol abuse, i.e. excessive acute alcohol use, combined with high dysfunctional impulsivity and antisocial behavior (Eensoo et al., 2003), or with personality traits such as impulsiveness, extroversion, sensation seeking, or novelty seeking in alcoholic subjects (Oreland, 2004). The discrepancies between results might be due to the different time period elapsed from the acute intoxication and sampling, since platelet MAO in our study was determined in sober alcohol-dependent subjects during withdrawal.

4.6. Platelet MAO Activity in Male and Female Alcoholic Patients and in Healthy Subjects

Platelet MAO activity is a biomarker that is affected by sex and smoking (Fowler et al., 2003; Oreland, 2004; Pivac et al., 2005), hence in our study we controlled these variables. Sex-related differences in platelet MAO activity were found, and the activity of the MAO in platelets was significantly increased in female subjects, both in alcoholic patients and in healthy women, when compared to the corresponding male subjects. Sex differences in platelet MAO activity, i.e. higher values of MAO in platelets in female than in male subjects, found in the present study, agree with our previous (Muck-Seler et al., 1991; Pivac et al., 2005) and most (Schooler et al., 1978; Verkes et al., 1998; Snell et al., 2002; Oreland, 2004; Coccini et al., 2005), but not all (Yates et al., 1990; Farren et al., 1998) reports. The reports who did not detect significant sex related differences in platelet MAO activity (Yates et al., 1990; Farren et al., 1998) had smaller number of subjects, and the results might be compromised with the possible effect of poly-drug use and alcohol related social complications, and/or longer (average 30 days) abstinent period (Yates et al., 1990), than in our study. The possible effect of drugs, that could alter the values of platelet MAO, was excluded, since our alcoholics were drug free. Sex-related differences in platelet MAO activity in our study are supported by the gender related differences found in 5-HT transporter gene, whole blood 5-HT level, platelet 5-HT binding in families with alcoholism (Stoltenberg et al., 2002), and in platelet 5-HT concentration in alcoholic (present study, Pivac et al., 2004) and control subjects (present study, Muck-Seler et al., 1996a; 1999; Pivac et al., 2001; 2004; 2005). Female subjects might have elevated activity of MAO in platelets because of the higher platelet MAO-B protein concentration in women (Snell et al., 2002). Platelet MAO activity might presumably be affected by the different phases of menstrual cycle (Leibenluft et al., 1994), and the opposite effects of estrogens on platelet MAO activity have been reported, with decline and/or peak of platelet MAO activity during ovulation (Leibenluft et al., 1994). However, in contrast to this study, it has been shown that platelet MAO activity did not vary during the menstrual cycle in control women (Hallman et al., 1987). Although in our study alcoholic women were significantly older than healthy women, and therefore some were in pre-menopausal and some in post-menopausal period, they all had almost identical values of platelet MAO activity, that were increased compared to male corresponding subjects, hence the possible effect of estrogens on platelet MAO activity might be neglected.

4.7. Platelet MAO Activity in Male and Female Alcoholic Patients and in Healthy Subjects Subdivided into Smokers and Nonsmokers

Since platelet MAO activity is affected by the substances from the cigarette smoke (Oreland et al., 2002; Oreland, 2004; Fowler et al., 2003; Pivac et al., 2006c), we divided alcoholic patients according to their smoking status. Our data showed only a trend towards the higher platelet MAO activity in nonsmokers compared to smokers within male and female alcoholic patients. On the other hand, smoking had a clear decreasing effect on platelet MAO activity in healthy men and women, and platelet MAO activity was significantly increased in healthy nonsmoking women when compared to female smokers, or to male smokers or nonsmokers. Nonsmoking healthy men had significantly higher platelet MAO activity than the corresponding men who smoked.

The reduction in platelet MAO activity in a dose-dependent manner has been shown to correlate with the number of cigarettes smoked per day (Eensoo et al., 2004). Since we have detected significant differences in platelet MAO activity between smokers and nonsmokers, we further evaluated the relationship between number of cigarettes smoked per day and platelet MAO activity. Although before (Pivac et al., 2005) we did not detect a significant difference in platelet MAO activity between alcoholic smokers who smoked 20, 40 or 60 cigarettes a day, in the present study we have substantially enlarged the groups of alcoholic patients and healthy control subjects and we included the data on the number of cigarettes smoked per day also for the healthy control subjects. Platelet MAO activity showed a significant fall that was conversely related to number of cigarettes smoked per day. Male healthy nonsmokers had the highest MAO activity, which declined significantly according to the number of cigarettes smoked per day, and this fall showed a dose-dependent pattern: it was moderate in smokers who smoked 20 cigarettes per day, and significant when healthy men smoked 40 cigarettes per day, and there was a trend towards the fall in healthy men who smoked 60 cigarettes per day (which did not reach the level of significance due to only 3 subjects). In addition, female subjects who smoked 20 cigarettes per day had significantly lower platelet MAO activity than healthy nonsmoking women. A trend of the reduced platelet MAO activity that was related to the number of cigarettes smoked per day, was observed in alcoholic male or female patients, but this trend, i.e. low/lower/the lowest platelet MAO activity in alcoholics who smoked 20, 40 or 60 cigarettes per day, respectively, compared to nonsmoking alcoholics, did not reach the level of statistical significance. Our data do not support the data from the intoxicated drunken drivers (Eensoo et al., 2004), however we are showing a pattern since we have found significant decreasing effect of the number of cigarettes smoked per day on platelet MAO activity only in healthy subjects. These data were corroborated with a detailed statistical evaluation (two-way ANOVA), that revealed that significant effect of smoking only within healthy, but not within alcoholic subjects. This result is in line with the similar values of platelet MAO levels between alcoholic smokers and non-smokers (Farren et al., 1998), but in contrast with lower MAO-B activity in smokers when compared to nonsmokers during alcohol intoxication, and after 8 days of abstinence (Schmidt et al., 1997). These differences might be explained by the inclusion of different diagnostic criteria, i.e. acutely intoxicated patients, abstinent alcoholics (Schmidt et al., 1997) vs. alcohol dependent patients in withdrawal, as in our study.

Smoking contributes to a lower MAO activity (Fowler et al., 2003; Pivac et al., 2005, 2006c), active substances from the cigarette smoke might be associated with development of addiction (Fowler et al., 2003), and nicotine releases the brain dopamine, noradrenalin or 5-HT, and inhibits the activity of MAO-A and MAO-B (Fowler et al., 2003), linking smoking with other forms of addictions, and/or psychiatric disorders (Berlin et al., 1995). Smoking presents a major problem in the public health, it carries a major risk for mental health, and patients with different psychiatric disorders are often heavy smokers.

4.8. Platelet MAO Activity in Male and Female Alcoholic Patients and in Healthy Subjects Subdivided According to the Different Psychiatric Comorbidities

Since alterations in platelet MAO were associated with different psychopathological conditions (Oreland, 2004), we determined platelet MAO in alcoholics with or without various comorbid psychiatric diagnoses (personality disorder, PTSD, anxious-depressive disorder, major depression, schizophrenia). The comorbid psychiatric entities did not influence platelet MAO activity within male or female alcoholic patients subdivided into those with or without psychiatric comorbidities. In the present study there was no significant effect between different comorbid diagnoses and platelet MAO activity, and no interaction between them, in alcoholic subjects. Our data show that male alcoholics with comorbid personality disorder, anxious-depressive disorder, depression, PTSD, schizophrenia, had similar values of platelet MAO activity as male alcoholic patients without these psychiatric disorders, or than patients with these psychiatric diagnoses who were not alcoholics. This finding is in line with similar platelet MAO values in nonalcoholic male war veterans with or without PTSD (Pivac et al., 2002), or between PTSD patients with a positive or negative family history of alcohol abuse (Davidson et al., 1985). Likewise, platelet MAO activity did not differ significantly within female alcoholics subdivided according to the presence or absence of comorbid personality disorder, anxious-depressive disorder, or depression. These data agree with the similar platelet MAO activity between nonalcoholic female depressed patients and healthy control women (Muck-Seler et al., 2002; Pivac et al., 2003), or between smokers with or without past history of depression (Berlin et al., 1995). The data suggest that platelet MAO is not affected by different psychiatric disorders comorbid with alcoholism, a finding in agreement with no significant association between depression, panic disorder, social phobia or serious attempts of suicide in alcohol-dependent subjects (Whitfield et al., 2000). The results are in agreement with the proposed role of platelet MAO activity as a trait dependent factor for the increased vulnerability to particular psychopathology and personality, and not for the psychiatric disorders (Farren et al., 1998). Male alcoholics, contrarily to female alcoholics, had comorbid schizophrenia and PTSD, suggesting that male alcoholics had a more severe alcohol dependency than female alcoholics, as detected previously (King et al., 2003).

It has been reported that platelet MAO increases with age (Veral et al., 1997; Oreland, 2004; Nicotra et al., 2004), and also that age does not affect platelet MAO (Coccini et al., 2005). In line with the lack of effect of older age on platelet MAO activity (Coccini et al., 2005; Pivac et al., 2005, 2006c), we did not find any significant correlation between platelet MAO activity and age of the subjects, either in smokers or in nonsmokers. In our study male and female alcoholics were older than healthy controls, but platelet MAO was not increased in alcoholics when compared to healthy controls. In addition, in line with previous reports (Farren et al., 1998; Pivac et al., 2005; 2006c), no association was found between platelet MAO activity and age in male or female alcoholics, or between male or female healthy subjects, therefore the possible influence of age on platelet MAO activity might be excluded. To avoid the influence of ethnicity on platelet MAO activity (Costa-Mallen et al., 2005), present study included only Caucasian subjects from Croatia.

Alcohol biomarkers, used to detect ethanol consumption are ethyl glucuronide, phosphatidyl ethanol, and ethyl sulfate. Biomarkers are very important research tool in alcoholism, since they can be used to determine the baseline characteristics of the alcoholic groups, to predict a suicidal risk in alcoholism (Gorwood, 2001), to predict efficacy in alcoholism treatment trials, to predict therapeutic response to medications in alcoholism, to measure alcohol consumption, and to be used in forensic toxicology. Although platelet MAO has been reported to be reduced in alcoholism (Oreland et al., 2002; Oreland, 2004), our data showed increased platelet MAO activity in male and unaltered platelet MAO activity in female alcoholic patients. Therefore, our data do not support the hypothesis that platelet MAO might be used as a trait marker of alcoholism, at least not in our group of drug-free alcoholics with withdrawal syndrome related to alcohol dependency. The alterations in platelet MAO activity were consistently found in various groups with different personality and temperamental traits, and therefore platelet MAO might serve as a trait marker associated with vulnerability to some personality traits such as sensation seeking and impulsivity (Oreland et al., 2002), impulse and affect dysregulation (Verkes et al., 1998), or social mal-adaptation (Schalling et al., 1987; Oreland et al., 2002; Longato-Stadler et al., 2002; Oreland, 2004), or biological marker of psychopathy, related to common personality factors among criminal recidivists (Stalenheim, 2004), but not for the alcoholism per se.

Our data showed different platelet MAO activity in male and female alcoholic subjects when compared to the corresponding control men or women. Platelet MAO activity in alcoholics was not affected by various psychiatric comorbid diagnoses, smoking status, or the number of smoked cigarettes per day. Pronounced sex-related difference in platelet MAO activity, both in alcoholic and healthy subjects, were found. Women in alcoholic and control groups had significantly higher platelet MAO activity than the corresponding male subjects. The decreasing effect of tobacco smoke on platelet MAO activity was significant only on healthy control subjects. Our data question the hypothesis of a reduced platelet MAO activity as a possible biological marker for alcoholism.

Chapter 5

Conclusion

The results from our present study, showing significant differences in platelet 5-HT markers in alcoholic patients, add further support to the role of 5-HT in the pathophysiology of alcoholism, since 5-HT alterations contribute to aggression (Goveas et al., 2004), selective serotonin reuptake inhibitors reduce alcohol intake, reduce the number of drinks, reduce the desire to drink, and increase the number of abstinent days (LeMarquand et al., 1994).

Regarding platelet MAO activity, our data do not support the hypothesis that platelet MAO might be used as a trait marker of alcoholism, at least not in our group of drug-free alcoholics with withdrawal syndrome related to alcohol dependency.

Acknowledgement

This work was supported by Croatian Ministry of Science, Education and Sport, Grant numbers 0098088 and 098-0982522-2455.

References

[1] American Psychiatric Association (1994). Diagnostic and Statistical Manual of Mental Disorders. IV Edition. *American Psychiatric Press*, Washington, DC.

[2] Alvarez, J.C., Cremniter, D., Lesieur, P., Gregoire, A., Gilton, A., Macquin-Mavier, I. (1999). Low blood cholesterol and low platelet serotonin levels in violent suicide attempters. *Biological Psychiatry 45*, 1066-1069.

[3] Andres, A.H., Rao, M.A., Ostrowitzki, S. and Enzian, W. (1993). Human brain cortex and platelet $serotonin_2$ receptor binding properties and their regulation by endogenous serotonin. *Life Sciences 52*, 313-321.

[4] Anthenelli, R.M., Tipp, J., Li, T-K., Magnes, L., Schuckit, M.A., Rice, J., Daw, W. and Nurnberger J.I. Jr. (1998). Platelet monoamine oxidase activity in subgroups of alcoholics and controls: Results from the collaborative study on the genetics of alcoholism. *Alcoholism: Clinical and Experimental Research 22,* 598-604.

[5] Arranz, B., Rosel, P., Sarro, S., Zaldivar, E., Cano, R., Navarro, M. and San, L. (1999). Platelet serotonergic binding sites in alcohol-dependent patients. *Alcohol and Alcoholism 34*, 726–732.

[6] Asberg, M., Traskman, L., Thoren, P. (1976). 5-HIAA in the cerebrospinal fluid. A biochemical suicide predictor? *Archives of General Psychiatry 33*, 1193–1197.

[7] Askenazy, F., Caci, H., Myquel, M., Darcourt, G. and Lecrubier, Y. (2002). Relationship between impulsivity and platelet serotonin content in adolescents. *Psychiatry Research* 94, 19-28.

[8] Bailly, D., Vignau, J., Lauth, B., Racadot, N., Beuscart, R., Servant, D. and Parquet, P.J. (1990). Platelet serotonin decrease in alcoholic patients. *Acta Psychiatrica Scandinavica 81,* 68-72.

[9] Bailly, D., Vignau, J., Racadot, N., Beuscart, R., Servant, D. and Parquet P.J. (1993). Platelet serotonin levels in alcoholic patients: Changes related to physiological and pathological factors. *Psychiatry Research 47*, 57-88.

[10] Bailey, C., Andrews, N., McKnight, A., Hughes, J. and Little, H. (2000). Prolonged changes in neurochemistry of dopamine neurons after chronic ethanol consumption. *Pharmacology Biochemistry and Behavior 66*, 153–161.

[11] Ballenger, J., Goodwin, F. and Major L. (1979). Alcohol and central serotonin metabolism in man. *Archives of General Psychiatry* 36, 224–227.

[12] Banki, C.M. (1978). 5-Hydroxytryptamine content of whole blood in psychiatric illness and alcoholism. *Acta Psychiatrica Scandinavica 57*, 232–238.

[13] Banki, C.M. (1981). Factors influencing monoamine metabolites and tryptophan in patients with alcohol dependence. *Journal of Neural Transmission 50*, 89–101.

[14] Balldin, J., Berggren, U., Engel, J. and Eriksson, M. (1994). Neuroendocrine evidence for reduced serotonergic neurotransmission during heavy drinking. *Alcoholism: Clinical and Experimental Research 18,* 822–825.

[15] Ballenger, J., Goodwin, F., Major, L. and Brown, G. (1979). Alcohol and central serotonin metabolism in man. *Archives of General Psychiatry 36*, 224–227.

[16] Berlin, I., Said, S., Spreux-Varoquaux, O., Olivares, R., Launay, J.M. and Puech, A.J. (1995). Monoamine oxidase A and B activities in heavy smokers. *Biological Psychiatry 38*, 756-761.

[17] Billett, E.E. (2004). Monoamine Oxidase (MAO) in human peripheral tissues. *Neurotoxicology 25*, 139-148.

[18] Blum, I., Vered, Y., Lifshitz, A., Harel, D., Blum, M., Nordenberg, Y., Harsad, A., Sulkes, J., Gabbay, U. and Graff, E. (1996). The effect of estrogen replacement therapy on plasma serotonin and catecholamines in postmenausal women. *Israel Journal of Medical Sciences 32*, 1158-1162.

[19] Branchey, L., Shaw, S. and Lieber, C.S. (1981). Ethanol impairs tryptophan transport into the brain and depresses serotonin. *Life Sciences 29,* 2751–2755.

[20] Bondy, B., Buettner, A. and Zill P. (2006). Genetics of suicide. *Molecular Psychiatry 4*, 1–16.

[21] Braunig, P., Rao, M.L., Fimmers, R. (1989). Blood serotonin levels in suicidal schizophrenic patients. *Acta Psychiatrica Scandinavica 79*, 186-189.

[22] Borcsiczky, D., Szalay, F., Tekes, K., Tarcali, J., Magyar, K. and Chatel, R. (1996). Platelet serotonin (5-HT) content is decreased in patients with alcoholic liver chirrhosis, but elevated in Gilbert's syndrome. *Journal of Hepatology 25*, 781-782.

[23] Buchsbaum, M.S., Coursey, R.D. and Murphy, D.L. (1976). The biochemical high-risk paradigm: behavioural and familial correlates of low platelet monoamine oxidase activity. *Science 194,* 339–341.

[24] Buydens-Branchey, L., Branchey, M.H., Noumair, D. and Lieber, C. S. (1989). Age of alcoholism onset: II. Relationship to susceptibility to serotonin precursor availability. *Archives of General Psychiatry 46*, 231–236.

[25] Camacho, A. and Dimsdale, J.E. (2000). Platelets and psychiatry: lessons learned from old and new studies. *Psychosomatic Medicine 62*, 326-336.

[26] Chen, H.-T., Casanova, M.F., Kleinman, J.E., Zito, M., Goldman, D. and Linnoila, M. (1991). ^{3}H-Paroxetine binding in brains of alcoholics. *Psychiatry Research* 38, 293–299.

[27] Cherek, D.R., Moeller, F.G., Khan-Dawood, F., Swann, A. and Lane S.D. (1999). Prolactin response to buspirone was reduced in violent compared to nonviolent parolees. *Psychopharmacology 142,* 144–148.

[28] Cloninger, C. R. (1987). Neurogenetic adaptive mechanisms in alcoholism. *Science 236*, 410–416.

[29] Coccini, T., Randine, G., Castoldi, A.F., Balloni, L., Baiardi, P. and Manzo, L. (2005). Lymphocyte muscarinic receptors and platelet monoamine oxidase-B as biomarkers of CNS function: effects of age and gender in healthy humans. *Environmental Toxicology and Pharmacology 19,* 715-720.

[30] Coccaro, E.F., Kavoussi, R.J., Sheline, Y.I., Lish, J.D. and Csernansky, J.G. (1996). Impulsive aggression in personality disorder correlates with tritiated paroxetine binding in the platelet. *Archives of General Psychiatry 53,* 531-536.

[31] Coccaro, E.F., Siever, L.J., Klar, H.M., Maurer, G., Cochrane, K., Cooper, T.B., Mohs, R.C. and Davis K.L. (1989). Serotonergic studies of personality disorder: Correlates with behavioral aggression and impulsivity. *Archives of General Psychiatry 46,* 587–599.

[32] Costa-Mallen, P., Keleda, S.N., Costa L.G. and Checkoway, H., (2005). Characterisation of the in vitro transcriptional activity of polymorphic alleles of the human monoamine oxidase-B gene. *Neuroscience Letters 383*, 171-175.

[33] Cremniter, D., Thenault, M., Jamain, S., Meidinger, A., Delmas, C. and Gaillard, M., (1994). Serotonine and suicide: a preliminary study concerning a sample of violent suicidal patients. *Progress in Neuropsychopharmacology and Biological Psychiatry 18*, 871-878.

[34] Da Prada, M., Cesura, A.M., Launay, J.M. and Richards, J.G. (1988). Platelets as models for neurons? *Experimental Research 44*, 115–126.

[35] Davidson, J., Lipper, S., Kilts, C.D., Mahorney, S. and Hammett, E. (1985). Platelet MAO activity in posttraumatic stress disorder. *American Journal of Psychiatry 142*, 1341-1343.

[36] Devor, E.J., Cloninger, C.R., Hoffman, P.L. and Tabakoff, B., (1993). Association of monoamine oxidase (MAO) activity with alcoholism and alcoholic subtypes. *American Journal of Medical Genetics 48*, 209–213.

[37] Dillon, K.A., Gross-Isseroff, R., Israeli, M. and Biegon, A. (1991). Autoradiographic analysis of serotonin 5-HT_{1A} receptor binding in the human brain post mortem: effects of age and alcohol. *Brain Research 554*, 56–64

[38] Dinan, T.G. (1996). Serotonin and the regulation of hypothalamic-pituitary-adrenal axis function. *Life Sciences 58*, 1683-1694.

[39] Dubovsky, S.L. and Thomas, M. (1995). Beyond specificity: Effects of serotonin and serotonergic treatments on psychobiological dysfunction. *Journal of Psychosomatic Research 39*, 429-444.

[40] Edenberg, H.J. and Kranzler, H.R. (2005). The contribution of genetics to addiction therapy approaches. *Pharmacology and Therapeutics 108,* 86-93.

[41] Eensoo, D., Paaver, M., Pulver, A., Harro, M. and Harro, J. (2004). Low platelet MAO activity associated with high dysfunctional impulsivity and antisocial behavior: evidence from drunk drivers. *Psychopharmacology 172*, 356-358.

[42] Ekblom, J., Garpenstrand, H., Damberg, M., Chen, K., Shih, J.C. and Oreland, L. (1998). Transcription factor binding to the core promoter of the human monoamine oxidase B gene in the cerebral cortex and in blood cells. *Neuroscience Letters 258*, 101-104.

[43] Erspamer, V. and Asero, B. (1952). Identification of enteramine, specific hormone of enterochromaffin cells, as 5-hydroxytryptamine. *Nature 169*, 800-801.

[44] Ewing, J. A. (1984). Detecting alcoholism: the CAGE questionnaire. *Journal of the American Medical Association 252*, 1905–1907.

[45] Faraj, B.A., Olkowski, L. and Jackson, R.T. (1997). Prevalence of high serotonin uptake in lymphocytes of abstinent alcoholics. *Biochemical Pharmacology 53,* 53-57.

[46] Farren, C.K., Clare, A.W., Tipton, K.F. and Dinan, T.G. (1998). Platelet MAO activity in subtypes of alcoholics and controls in a homogenous population. *Journal of Psychiatric Research 32*, 49-54.

[47] Fils-Aime, M.L., Eckardt, M.J., George, D.T., Beown, D.L., Mefford, I. and Linnoila, M. (1996). Early onset alcoholics have lower cerebrospinal fluid 5HIAA levels than late onset alcoholics. *Archieves in General Psychiatry 53(3)*, 211-216.

[48] Fowler, J.S., Logan, J., Wang, G.J. and Volkow, N.D. (2003). Monoamine oxidase and cigarette smoking. *Neurotoxicology 24*, 75-82.

[49] Franke, L., Schewe, H.-J., Müller, B., Campman, V., Kitzrow, W., Uebelhack, R., Berghöfer, A. and Müller-Oerlinghausen, B. (2000). Serotonergic platelet variables in unmedicated patients suffering from major depression and healthy subjects. Relationship between 5HT content and 5HT uptake. *Life Sciences 47*, 301-315.

[50] Garpenstrand, H., Ekblom, J., Forslund, K., Rylander, G. and Oreland, L. (2000). Platelet monoamine oxidase activity is related to MAOB intron 13 genotype. *Journal of Neural Transmission 107*, 523-530.

[51] Goldman, D. (1995). Identifying alcoholism vulnerability alleles. *Alcoholism: Clinical and Experimental Research 19*, 824–831.

[52] Gorwood, P. (2001). Biological markers for suicidal behavior in alcohol dependence. *Eurpean Psychiatry 16*, 410-417.

[53] Goveas, J.S., Csernansky, J.G. and Coccaro, E.F. (2004). Platelet serotonin content correlates with life history of aggression in personality-disordered subjects. *Psychiatry Research* 126, 23-32.

[54] Gross-Isseroff, R. and Biegon, A. (1988). Autoradiographic analysis of [^{3}H]imipramine binding in the human brain post-mortem: effects of age and alcohol. *Journal of Neurochemistry 51*, 528–534.

[55] Gruenewald, P., Ponicki, W. and Mitchell, P. (1995). Suicide rates and alcohol consumption in the United States, 1970-89. *Addiction 90*, 1063-1075.

[56] Guicheney, P., Leger, D., Barrat, J., Trevoux, R., De Lignieres, B., Roques, P., Garnier, J.P., Boyers, P., Grenier, J., Dreux, C. and Meyer, P. (1988). Platelet serotonin content and plasma tryptophan in peri- and postmenopausal women: variations with plasma oestrogen levels and depressive symptoms. *European Journal of Clinical Investigation 18,* 297-304.

[57] Hallman, J., Oreland, L., Edman, G. and Schalling, D. (1987). Thrombocyte monoamine oxidase activity and presonality traits in women with severe premenstrual syndrome. *Acta Psychiatrica Scandinavica 76,* 225-234.

[58] Hamilton, M. (1960). A rating scale for depression. *Journal of Neurology, Neurosurgery and Psychiatry 23*, 56-62.

[59] Heinz, A., Mann, K., Weinberger, D.R. and Goldman, D. (2001). Serotonergic dysfunction, negative mood states, and response to alcohol. *Alcoholism: Clinical and Experimental Research 25(4)*, 487-495.

[60] Hindberg, I. and Naesh, O. (1992). Serotonin concentrations in plasma and variations during menstrual cycle. *Clinical Chemistry 38*, 2087–2089.

[61] Hrdina, P.D. (1994) Platelet serotonergic markers in psychiatric disorders: use, abuse and limitations. *Journal of Psychiatry and Neurosciences 19*, 87-88.

[62] Jakovljevic, M., Muck-Seler, D., Pivac, N., Ljubicic, Đ., Bujas, M. and Dodig, G., (1997). Seasonal influence on platelet 5-HT concentrations in patients with recurrent major depression and schizophrenia. *Biological Psychiatry 41*, 1028-1034.

[63] Javors, M., Tiouririne, M. and Prihoda, T. (2000). Platelet serotonin uptake is higher in early-onset than in late-onset alcoholics. *Alcohol and Alcoholism 35(4)*, 390-393.

[64] Johnson, B.A. (2004). Role of the serotonergic system in the neurobiology of alcoholism - Implications for treatment. *CNS Drugs 18*, 1105-1118.

[65] Johnson, B.A. and Ait-Daoud, N. (2000). Neuropharmacological treatments for alcoholism: scientific basis and clinical findings. *Psychopharmacology 149*, 327-344.

[66] Kema, I.P., de Vries, E.G.E. and Muskiet, F.A.J. (2000). Clinical chemistry of serotonin and metabolites. *Journal of Chromatography B 747,* 33-48.

[67] Kent, T.A., Campbell, J.L., Pazdernik, T.L., Hunter, R., Gunn, W.H. and Goodwin, D.W. (1985). Blood platelet uptake of serotonin in men alcoholics. *Journal of the Studies on Alcohol 46(4)*, 357-359.

[68] King, A.C., Bernardy, N.C. and Hauner, K. (2003). Stressful events, personality, and mood disturbance: gender differences in alcoholic and problem drinkers. *Addictive behaviors 28*, 171-187.

[69] Kozaric-Kovacic, D., Folnegovic-Smalc, V., Jakovljevic, M. and Lang, B. (1992). Social trauma and stress in alcoholic patients. *Alcoholism 28*, 85-90.

[70] Kozaric-Kovacic, D., Folnegovic-Smalc, V., Folnegovic, Z. and Marusic, A. (1995). Influence of alcoholism on the prognosis of schizophrenic patients. *Journal of studies on alcohol 56*, 622-627.

[71] Kozaric-Kovacic, D., Ljubin, T. and Grappe, M. (2000). Comorbidity of posttraumatic stress disorder and alcohol dependence in displaced persons. *Croatian medical journal 41*, 173-178.

[72] Kozaric-Kovacic, D., Ljubin, T., Rutic-Puz, L., Mihaljevic, J., Jernej, B. and Cicin-Sajn, L. (2000). Platelet monoamine oxidase activity, ego strength, and neuroticism in soldiers with combat related current posttraumatic stress disorder. *Croatian Medical Journal 41*, 76-80.

[73] Krajl, M. (1965). A rapid microfluorimetric determination of monoamine oxidase. *Biochemical Pharmacology 14*, 1683-1685.

[74] Krystal, J.H., Webb, E., Cooney, N.L., Kranzler, H.R. and Charney, D.S. (1994). Specificity of ethanol-like effects elicited by serotonergic and noradrenergic mechanisms. *Archives of General Psychiatry 51*, 898–911.

[75] Krystal, J.H., Webb, E., Cooney, N.L., Kranzler, H.R., Southwick, S.W., Heninger, G.L. and Charney, D.S. (1996). Serotonergic and noradrenergic dysregulation in alcoholism: m-chlorophenylpiperazide and yohimbine effects in recently detoxified alcoholics and healthy subjects. *American Journal of Psychiatry 153*, 83–92.

[76] Kumar, A.D., Weiss, S., Fernandez, J.B., Cruess, D. and Eisdorfer, C. (1998). Peripheral serotonin levels in women: role of aging and ethnicity. *Gerontology 44*, 211-216.

[77] Lee, M.A. and Meltzer, H.Y. (2001). 5-HT1A receptor dysfunction in female patients with schizophrenia. *Biological Psychiatry 50*, 758-766.

[78] Leibenluft, E., Fiero, P.L. and , David, R. (1994). Effects of menstrual cycle on dependent variables in mood disorder research. *Archives in General Psychiatry 51*, 761-781.

[79] LeMarquand, D., Pihl, R.O. and Benkelfat, C. (1994). Serotonin and alcohol intake, abuse, and dependence: Clinical evidence. *Biological Psychiatry 36(6)*, 326-337.

[80] Lesch, K.P., Baling, U., Gross, J., Strauss, K., Wolozin, B., Murphy, D.L. and Riederer, P. (1994). Organization of the human serotonin transporter gene. *Journal of Neuronal Transmission 95*, 157-162.

[81] Leyton, M., Corin, E., Martial, J. and Meany, M. (1995). Psychotic symptoms and vulnerability to recurrent major depression. *Journal of Affective Disorders 33*, 107-115.

[82] Linnoila, M., Virkkunen, M., Scheinin, M., Nuutila, A., Rimon, R. and Goodwin, F. K. (1983). Low cerebrospinal fluid 5-hydroxyindoleacetic acid concentration differentiates impulsive from nonimpulsive violent behavior. *Life Sciences 33*, 2609–2614.

[83] Longato-Stadler, E., Klinteberg, B., Garpenstrand, H., Oreland, L. and Hallman, J. (2002). Personality traits and platelet monoamine oxidase activity in a Swedish male criminal population. *Neuropsychobiology 46*, 202-208.

[84] Lovinger, D.M. (1997). Serotonin's role in alcohol's effects on the brain. *Alcohol health and research world 21*, 114-120.

[85] Lowry, O.H., Rosenbrough, N.S., Farr, A.C. and Randall, R.J. (1951). Protein measurement with the Folin phenol reagent. *Journal of Biological Chemistry 193*, 265-75.

[86] Lucki, I. (1998). The spectrum of behaviors influenced by serotonin. *Biological Psychiatry 44*, 151-162.

[87] Mann, J.J., (1998). The neurobiology of suicide. *Nature medicine 4*, 25–30.

[88] Mann, J.J., and Malone, K.M. (1997). Cerebrospinal fluid amines and higher lethality suicide attempts in depressed inpatients. *Biological psychiatry 41*, 162–171.

[89] Mann, J.J., Henteleff, R.A., Lagattuta, T.F., Perper, J.A., Li, S. and Arango, V. (1996). Lower 3H-paroxetine binding in cerebral cortex of suicide victims is partly due to fewer high affinity, non-transporter sites. *Journal of neural transmission 103*, 1337–1350.

[90] Mattson, M.P., Maudsley, S. and Martin, B. (2004). BDNF and 5-HT: a dynamic duo in age-related neuronal plasticity and neurodegenerative disorders. *Trends in Neurosciences 27*, 589-594.

[91] Mendelson, S.C. (2000). The current status of the platelet 5-HT2A receptor in depression. *Journal of Affective Disorders 57*, 13-24.

[92] Meszaros, Z., Borsiczky, D., Mate, M., Tarcali, J., Szombathy, T., Tekes, K. and Magyar, K. (1998). Platelet MAO-B activity and serotonin content in patients with dementia: effect of age, medication and disease. *Neurochemical research 23*, 863-868.

[93] Moeller, F.G. and Dougherty, D.M. (2001). Antisocial personality disorder, alcohol, and aggression. *Alcohol research and health : the journal of the National Institute on Alcohol Abuse and Alcoholism 25*, 5-11.

[94] Mezzich, A.C., Tarter, R.E., Moss, H.B., Yao, J.K., Hsieh, Y-C. and Kirisci, L. (1994), Platelet monoamine oxidase activity and temperament and personality in adolescent female substance abusers. Personality and individual differences *16*, 417-424.

[95] Moore, T., Scarpa, A. and Raine, A. (2002). A meta-analysis of serotonin metabolite 5-HIAA and antisocial behavior. *Aggressive behavior 28*, 299–316.

[96] Muck-Seler, D., Jakovljevic, M. and Deanovic, Z. (1991). Platelet serotonin in subtypes of schizophrenia and unipolar depression. *Psychiatry Research 38*, 105-13.

[97] Muck-Seler, D., Pivac, N. and Jakovljevic, M. (1999). Sex differences, season of birth and platelet 5-HT levels in schizophrenic patients. *Journal of Neural Transmission 106,* 337-347.

[98] Muck-Seler, D., Jakovljevic, M. and Pivac, N. (1996a). Platelet 5-HT concentrations and suicidal behavior in recurrent major depression. *Journal of Affective Disorders 39*, 73-80.

[99] Muck-Seler, D., Bujas, M., Ljubić-Thibal, V. and Jakovljevic, M. (1996b). Effect of age on platelet 5-HT concentrations in healthy controls, depressed and schizophrenic patients. *Neuropsychobiology 36*, 19-21.

[100] Muck-Seler, D., Pivac, N., Sagud, M., Jakovljevic, M. and Mihaljević-Peleš, A. (2002). The effects of paroxetine and tianeptine on peripheral biochemical markers in major depression. *Progress in Neuro-Psychopharmacology and Biological Psychiatry 26*, 1235-1243.

[101] Muck-Šeler, D., Pivac, N., Jakovljević, M., Šagud, M. and Mihaljević-Peleš, A. (2003). Platelet 5-HT concentration and comorbid depression in war veterans with or without posttraumatic stress disorder. *Journal of Affective Disorders 75*, 171-179.

[102] Muck-Seler, D., Pivac, N., Mustapic, M., Crncevic, Z., Jakovljevic, M. and Sagud, M. (2004). Platelet serotonin and plasma prolactin and cortisol in healthy, depressed and schizophrenic women. *Psychiatry Research 127*, 217-226.

[103] Muck-Seler, D., Pivac, N., Sagud, M., Mustapic, M and Jakovljevic, M. (2005). The effects of serotonin uptake inhibitors on platelet serotonin: From basic to clinical studies. In Shirley AC (Ed.), *Trends in serotonin uptake inhibitors research*, NOVA Science Publishers, Inc, NY, Hauppauge, USA, 29-53.

[104] Muller-Oerlinghausen, B., Roggenbach, J. and Franke, L. (2004). Serotonergic platelet markers of suicidal behavior – do they really exist? *Journal of Affective Disorders 79*, 13-24.

[105] Nicotra, A., Pierucci, F., Parvez, H. and Senatori, O. (2004). Monoamine oxidase expression during development and aging. *Neurotoxicology 25*, 155-165.

[106] Nishizawa, S., Benkelfat, C., Young, S.N., Leyton, M., Mzengeza, S., De Montigny, C., Blier, P. and Diksic, M. (1997). Differences between males and females in rates of serotonin synthesis in human brain. *Proceedings of the National Academy of Sciences of USA 94*, 5308-5313.

[107] Oreland, L. and Shaskan, E.G. (1983). Monoamine oxidase activity as a biological marker. *Trends in Pharmacological Sciences 4*, 339-341.

[108] Oreland, L. and Hallman, J. (1995). The correlation between platelet MAO activity and personality—a short review of findings and a discussion on possible mechanisms. In: Yu PU et al. (Eds.), *Progress in Brain Research, vol 106. Current Neurochemical and Pharmacological Aspects of Biogenic Amines: Their Function, Oxidative Deamination and Inhibition.* Elsevier, New York, 77–84.

[109] Oreland, L., Damberg, M., Hallman, J. and Garpenstrand, H. (2002). Smoking only explains part of the associations between platelet monoamine oxidase activity and personality. *Journal of Neural Transmission 109*, 963-975.

[110] Oreland, L. (2004). Platelet monoamine oxidase, personality and alcoholism: The rise, fall and resurrection. *Neurotoxicology 25*, 79-89.

[111] Oxenkrug, G.F. (1979). The content and uptake of 5-HT by blood platelets in depressive patients. *Journal of Neural Transmission 45*, 285-289.

[112] Patkar, A.A., Gopalakrishnan, R., Naik, P.C., Murray, H.W., Vergare, M.J. and Marsden, C.A. (2003). Changes in plasma noradrenaline and serotonin levels and craving durimg alcohol withdrawal. *Alcohol and Alcoholism 38*, 224-231.

[113] Pivac, N., Jakovljevic, M., Muck-Seler, D., and Brzović, Z. (1997a). Hypothalamic-pituitary-adrenal axis function and platelet serotonin concentrations in depressed patients. *Psychiatry Research 73,* 123-132.

[114] Pivac, N., Muck-Seler, D. and Jakovljević, M. (1997b). Platelet 5-HT levels and hypothalamic-pituitary-adrenal axis activity in schizophrenic patients with positive and negative symptoms. *Neuropsychobiology* 36, 19-21.

[115] Pivac, N., Mück-Šeler, D., Barišić, I., Jakovljević, M., and Puretić, Z. (2001). Platelet serotonin concentration in dialysis patients with somatic symptoms of depression. *Life Sciences 68*, 2423-2433.

[116] Pivac, N., Muck-Seler, D., Sagud, M. and Jakovljevic, M. (2002). Platelet serotonergic markers in posttraumatic stress disorder. *Progress in Neuro-Psychopharmacology and Biological Psychiatry* 26, 1193-1198.

[117] Pivac, N., Muck-Seler, D., Šagud, M., Jakovljevic, M., Mustapić, M., and Mihaljević-Peleš, A. (2003). Long-term sertraline treatment and peripheral biochemical markers in female depressed patients. *Progress in Neuro-Psychopharmacology and Biological Psychiatry 27*, 759-765.

[118] Pivac, N., Muck-Seler, D., Mustapic, M., Nenadic-Sviglin, K. and Kozaric-Kovacic, D. (2004). Platelet serotonin concentration in alcoholic subjects. *Life Sciences 76*, 521-531.

[119] Pivac, N., Muck-Seler, D., Kozaric-Kovacic, D., Mustapic, M., Nenadic-Sviglin, K. and Dezeljin M. (2005). Platelet monoamine oxidase in alcoholism. Letter to the editors. *Psychopharmacolog 182,* 194-196.

[120] Pivac, N., Kozarić-Kovačić, D., Mustapic, M., Deželjin, M., Borovecki, A., Grubišic-Ilić, M. and Muck-Seler, D. (2006a). Platelet serotonin in combat related posttraumatic stress disorder with psychotic symptoms. *Journal of Affective Disorders*, doi: 10.1016/j.jad.2006.02.018.

[121] Pivac, N., Kozarić-Kovačić, D. and Mück-Šeler, D. (2006b). Biological markers in Croatian war veterans with combat related posttraumatic stress disorder. In: *Novel approaches to the diagnosis and treatment of posttraumatic stress disorder. NATO Security through Science series E: Human and Societal Dynamics – Vol 6. Section 1: Epidemiology and pathophysiology of PTSD.* Roy, M.J, ed. IOS Press Amsterdam Berlin Oxford Tokyo Washington DC, 3-12.

[122] Pivac, N., Knežević, J., Mustapic, M., Deželjin, M., Mück-Šeler D., Kozaric-Kovacic, D., Balija, M., Matijevic, T. and Pavelic, J. (2006c). The lack of association between monoamine oxidase (MAO) intron 13 polymorphism and platelet MAO activity among men. *Life Sciences 79,* 45-49.

[123] Ponzer, S., Johansson, S.E. and Bergman, B. (2002). A four year follow up study of male alcoholics: factors affecting the risk of readmission. *Alcohol 27*, 83-88.

[124] Popova, N.K. (2006). From genes to aggressive behavior: the role of serotonergic system. *Bioassays 28*, 495-503.

[125] Ratsma, J.E., Van Der Stelt, O. and Gunning, W.B. (2002). Neurochemical markers of alcoholism vulnerability in humans. *Alcohol and Alcoholism 37*, 522-533.

[126] Rao, M.L., Braunig, P. and Papassotiropoulos, A. (1994). Autoaggressive behavior is closely related to serotonin availability in schizoaffective disorder. *Pharmacopsychiatry 27,* 202-206.

[127] Rasgon, N., McGuire, M., Tanavoli, S., Fairbanks, L. and Rapkin, A. (2000). Neuroendocrine response to an intravenous L-tryptophan challenge in women with premenstrual syndrome. *Fertility and Sterility 73,* 144-149.

[128] Rosa-Neto, P., Diksic, M., Okazawa, H., Leyton, M., Ghadirian, N., Mzengeza, S., Nakai, A., Debonnel, A., Blier, P. and Benkelfat, C. (2004). Measurement of brain regional alpha -[^{11}C]-methyl-L-tryptophan trapping as a measure of serotonin synthesis in medication-free patients with major depression. *Archives of General Psychiatry 61*, 556-563.

[129] Roy, A. and Linnoila, M. (1988). Suicidal behavior, impulsiveness and serotonin. *Acta Psychiatrica Scandinavica 78,* 529–535.

[130] Sagud, M., Mihaljević-Peleš, A., Pivac, N., Jakovljević, M. and Mück-Šeler, D. (2006). Platelet serotonin and serum lipids in psychotic mania. *Journal of Affective Disorders*, accepted for publication.

[131] Schalling, D., Asberg, M., Edman, G. and Oreland, L. (1987). Markers for vulnerability to psychopathology: Temperament traits associated with platelet MAO activity. *Acta Psychiatrica Scandinavica 76 (2)*, 172-182.

[132] Schmidt, L.G., Dufeu, P., Heinz, A., Kuhn, S. and Rommelspacher H. (1997). Serotonergic dysfunction in addiction - effects of alcohol, cigarette smoking and heroin on platelet 5-HT content. *Psychiatry Research 72*, 177-185.

[133] Sher, L. (2006). Alcoholism and suicidal behavior: a clinical overview. *Acta Psychiatrica Scandinavica 113 (1)*, 13-22.

[134] Schulz, K.P., McKay, K.E., Newcorn, J.H., Sharma, V., Gabriel, S. and Halperin, J.M. (1998). Serotonin Function and Risk for Alcoholism in Boys with Attention-Deficit Hyperactivity Disorder. *Neuropsychopharmacology 18,* 10–17.

[135] Schooler, C., Zahn, T.P., Murphy, D.L. and Buschbaum, M.S. (1978). Psychological correlates of monoamine oxidase activity in normals. *Journal of Nervous and Mental Diseases 166,* 177-186.

[136] SAMHSA (2004). Results from the 2003 National Survey on Drug Use and Health (NSDUH Series H-25). Office of Applied Studies, Substance Abuse and Mental Health Services Administration, Rockville, MD.

[137] Snell, L.D., Glanz, J. and Tabakoff, B. (2002). Relationship between effects of smoking, gender, and alcohol dependence on platelet monoamine oxidase-B activity, affinity labeling, and protein measurements. *Alcoholism: Clinical and Experimental Research 26,* 1105-1113.

[138] Soloff, P.H., Lynch, K.G. and Moss, H.B. (2000). Serotonin, impulsivity, and alcohol use disorders in the older adolescent: A psychobiological study. *Alcoholism: Clinical and Experimental Research 24,* 1609-1619.

[139] Schuckit, M.A., Smith, T.L. and Jelger, K. (2004). Alcoholism, The Search for Genes Contributing to the Low Level of Response to Alcohol: Patterns of Findings Across Studies. *Alcoholism: Clinical and Experimental Research 28*, 1449-1458.

[140] Stahl, S.M. (1985). Platelets as pharmacological models for the receptors and biochemistry of monoaminergic neurons. In: Longenecker GL, (Ed.), *Platelets: Physiology and Pharmacology*, Academic Press, New York, 307-340.

[141] Stahl, S.M. (1998). Mechanism of action of serotonin selective reuptake inhibitors. Serotonin receptors and pathways mediate therapeutic effects and side effects. *Journal of Affective Disorders 51*, 215-35.

[142] Stalenheim, E. G. (2004). Long-term validity of biological markers of psychopathy and criminal recidivism: follow-up 6–8 years after forensic psychiatric investigation. *Psychiatry Research 121,* 281–291.

[143] Stoltenberg, S.F., Twitchell, G.R., Hanna, G.L., Cook, E.H., Fitzgerald, H.E., Zucker, R.A. and Little, K.Y. (2002). Serotonin transporter promoter polymorphism, peripheral indexes of serotonin function, and personality measures in families with alcoholism. *American Jurnal of Medical Genetics 114,* 230-234.

[144] Storvik, M., Tiihonen, J., Haukijärvi, T. and Tupala, E. (2006). Lower serotonin ransporter binding in caudate in alcoholics. *Synapse 59*, 144-151.

[145] Szalay, F., Borcsiczky, D., Kempler, P. and Tekes, K. (2001). Platelet serotonin (5-HT) is decreased not only in alcoholic but also in HCV cirrhosis. *American Journal of Gastroenterology 96*, Suppl, S137-S138.

[146] Twichell, G.R., Hanna, G.L., Cook, E.H., Fitzgerald, H.E., Little, K.Y. and Zucker, R.A. (1998). Overt behavior problems and serotonergic function in middle childhood among male and female offspring of alcoholic fathers. *Alcoholism: Clinical and Experimental Research 22,* 1340-1348.

[147] van Praag, H.M., (2004). The cognitive paradox in posttraumatic stress disorder: a hypothesis. *Progress in Neuro-Psychopharmacology and Biological Psychiatry 28*, 923-935.

[148] Veenstra-Vander Weele, J., Anderson, G.M. and Cook, E.H. Jr. (2000). Pharmacogenetics of the serotonin system: initial studies and future directions. *European Journal of Pharmacology 410,* 165-181.

[149] Veral, A., Alper, G., Mentes, G. and Ersoz, B. (1997). Age and sex related alterations in serum and platelet monoamine oxidase. *European Journal of Clinical Chemistry and Clinical Biochemistry 35,* 265-268.

[150] Verkes, R.J., Van der Mast, R.C., Kerkhof, A.J.F.M., Fekkes, D., Hengeveld, M.W., Tuyl, J.P. and Van Kempen, G.M.J. (1998). Platelet serotonin, monoamine oxidase activity, and [^{3}H] paroxetine binding related to impulsive suicide attempts and borderline personality disorder. *Biological Psychiatry 43,* 740-746.

[151] Virkkunen, M., Goldman, D., Nielsen, D.A. and Linnoila, M. (1995). Low brain serotonin turnover rate (low CSF 5-HIAA) and impulsive violence. *Journal of Psychiatry and Neurosciences 20*, 271–275.

[152] Volkow, N.D. and Li, T.H. (2005). Drugs and alcohol: Treating and preventing abuse, addiction and their medical consequences. *Pharmacology and Therapeutics 108*, 3-17.

[153] von Knorring, A.-L., Hallman, J., von Knorring, L. and Oreland, L. (1991). Platelet monoamine xidase activity in type 1 and type 2 alcoholism. *Alcohol and Alcoholism 26*, 409–416.

[154] Westenberg, H.G.M., Van Praag, H.M., de Jong, J.T.V.M. and Thijssen, J.H.H. (1982). Postsynaptic serotonergic activity in depressed patients: Evaluation of the neuroendocrine strategy. *Psychiatry Research 7*, 361-371.

[155] Whitfield, J.B., Pang, D., Bucholz, K.K., Madden, P.A., Heath, A.C., Statham, D.J. and Martin, N.G. (2000). Monoamine oxidase: associations with alcohol dependence, smoking and other measures of psychopathology. *Psychological Medicine 30*, 443-454.

[156] Winblad, B., Gottfries, C.G., Oreland, L. and Wiberg, A. (1979). Monoamine oxidase in platelets and brains of nonpsychiatric and non-neurological geriatric patients. *Medical Biology 57*, 129-132.

[157] Wirtz-Justice, A., Lichtsteiner, M. and Feer, H. (1977). Diurnal and seasonal variations in human platelet serotonin in man. *Journal of Neuronal Transmission 42*, 7-15.

[158] World Heath Organization (2002). The World Health report 2002: Reducing Risks, Promoting Heathy Life. Geneva.

[159] Yates, W.R., Wilcox, J., Knudson, R., Myers, C. and Kelly, M.W. (1990). The effect of gender and subtype on platelet MAO in alcoholism. *Journal of Studies on Alcohol 51*, 463-467.

[160] Zarcone, V., Barchas, J., Hoddes, E., Monplaisir, J., Sack, R. and Wilson, R. (1975). Experimental ethanol ingestion: Sleep variables and metabolites of dopamine and serotonin in the cerebrospinal fluid. *Advances in Experimental Medical Biology 59*, 431–451.

Index

D

E

F

G

H

N

O

P

Q

R

S